THE SPORTS VIDEO RESOURCE GUIDE

A Fan's Sourcebook for
All the Best in Sports Videos

Bob Carroll

A FIRESIDE BOOK
Published by Simon & Schuster
New York London Toronto Sydney Tokyo Singapore

FIRESIDE
Simon & Schuster Building
Rockefeller Center
1230 Avenue of the Americas
New York, New York 10020

Copyright © 1992 by Professional Ink, Inc. and Bob Carroll

All rights reserved
including the right of reproduction
in whole or in part in any form.

FIRESIDE and colophon are registered trademarks
of Simon & Schuster Inc.

Manufactured in the United States of America

1 3 5 7 9 10 8 6 4 2

Library of Congress Cataloging-in-Publication Data is available.

ISBN   0-671-73446-6

CONTENTS

Introduction 11
Baseball 19
    Listings 29
Basketball 58
    Listings 66
Football 80
    Listings 89
Hockey 124
    Listings 128
Soccer 133
    Listings 136
Boxing 140
    Listings 143
Tennis 147
    Listings 149
Golf 152
    Listings 154
General, Olympics, and Other Sports 177
    Listings 181
Movies 195
    Indexes 226

ACKNOWLEDGEMENTS

A lot of people helped put this book together. I can't list them all, but I'd particularly like to thank John McNamara for his help with the basketball section, John Thorn and Richard Puff of Professional Ink, Jonathan Hock and Kathy Davis of NFL Films, Paul Haas of Sports Books, etc., Bill Southard of Blockbuster, the helpful folks at CBS/Fox and Front Row Video, and Rich Killar and the gang at Alternative Ink, Ltd.

THE SPORTS VIDEO RESOURCE GUIDE

INTRODUCTION

Let's not kid each other.

Suppose I told you that I had watched every single video listed in this book. I could say I took copious notes, re-ran the choice sections over and over, sent away for scripts, and conferred with old men who had lists of letters after their names.

Most of you would believe it. After all, I never lied to you before, did I?

But a couple of you out there—the ones who have been burned by a smile, a shoeshine, and a pretty face on other occasions—would begin putting a few numbers together. Let's see, nearly 1,300 sports videos plus more than 120 sports movies.... How long would it take to watch all that?

Say I work a 40-hour week. Throw in a few holidays. Remember to allow for rewinding. Don't forget phone calls. Maybe a relative or friend drops by. Then allow two days for when the cat got sick. What's that come to?

At least ten months!

That's what I faced when I started figuring.

But I was willing. When it comes to dedication, hey! Count on me.

Alas! Then I got the bad news. The only way I could preview some of the videos on my list was to go out and buy them. Many of the companies that produce sports videos were as sweet as pie when it came to lending preview copies. And my local rental stores helped out. But try as I might, there are a few companies that take pay-per-view very seriously. Since I take my bank balance just as seriously, I finally faced the fact that I just wasn't going to see *every* video on my list.

So, like I said, let's not kid each other.

What I *did* do was watch all the sports videos I could. Find out everything I could about the ones I couldn't watch. And, in those cases where all I could learn from an uncommunicative producer was a title—well, that's what you get.

Actually, the most important thing is to make you aware of what's out there.

Except for movies, I won't rate one video against another because when you come right down to it, *you* have to decide what's best for you. I might find a certain highlight video boring, repetitive, and uninspired, but if it shows the best moments of *your* favorite team, you'll love it. I may think a particular golf instructional is simplistic and misses important points, but if it helps you to correct your slice, you'll want it bronzed. I admit to liking blooper videos in certain sports more than in others; your prejudice may be the opposite.

I tried to hit all the major sports, but not fitness tapes of the Jane Fonda variety. If they ever learn to score push-ups, I might reconsider. Likewise, I decided against professional wrestling. Those grunters and groaners are great athletes all right but so are ballet stars, and I'm not including *Swan Lake*. After all, unlike a ball game, the end of *Swan Lake* is known before it starts. Come to think of it, that applies to pro wrestling too.

What You Will Find Here

Title: It seems like nine out of ten sports videos have at least two titles. *Champions Against the Odds* and *1987 Denver Broncos* are the same tape. You'd best look for that highlight video under both titles, and there's no telling which will be listed first in somebody's catalog. In this book, I've tried hard to give full titles. But be careful when ordering. If you only have half a title, you may not get what you're looking for. *A New Beginning*, for example, is the subtitle of NFL Films' 1986 Green Bay highlight video AND its 1984 Pittsburgh video. In this book, each sport's chapter on videos concludes with an alphabetical index of titles.

Description: The descriptions of videos in this book are necessarily short. They're meant to help you find what you're seeking, not explain the video. When the title was self-explanatory, such as *Golden Greats of Tennis*, no description seemed necessary.

Production company / distributor: Suppose you've set your little ol' video-lovin' heart on owning *Monday Night Madness* because your buddy says it's great. The question is how to get it. Your pal tells you it's made by CBS/Fox Video, which happens to be in New York City. You get their phone number through directory information and give them a call. Guess what? They won't sell you a copy.

Okay, don't panic. Let's go down a step to distributors. You come across the name "Baker and Taylor." It's a big company with more than a dozen warehouses around the country. B & T distributes to video stores and also to sports equipment stores. Just what you need, right? Wrong! You're not a video or sports equipment store and they sell to retailers only.

Generally, you can't order a video directly from the label or distributor, so knowing the name of the company that made a particular sports video will do you about as much good as knowing the name of the farmer who grew the lettuce for your salad. All the same, I list most of the production companies or national distributors, anyway, because maybe, just maybe, that info will help your local video-store owner source the title for you in one of his catalogs.

A key to the producer/distributor codes will be found at the end of this Introduction.

Release date: The year a video came out can be useful in eliminating some of the confusion caused by similar titles. I gave you the year in most cases when I could learn it. But you can figure out the date of say, a Detroit Pistons' video titled DETROIT PISTONS 1987–88: *Bad Boys*. And listing the year of the video can add to confusion when dealing with an historical tape that was *filmed* thirty or forty years ago.

Color or black and white: Although movies come in black-and-white, color, or computer colorization, almost all sports videos taken from films made before the mid-1950s are in black-and-white. Nearly every post-1960 video is in color. But many historical videos mix color and black-and-white. All videos listed here are in color except for those that incorporate films from before 1960. When there was an exception, I listed it.

Running time: For most videos I also listed the running times, but frankly, those are not to be trusted. A lot of "30 minute" videos go 45 minutes, and so do some "60 minute" videos. Call them "ballpark figures."

Price: I listed either the suggested retail price or, more often, the lowest price I came across. But videos are like any other commodity. The longer they sit unsold on the shelves, the more likely the price will drop. Shop around.

Tapes are listed by sport with the only exception being our "General, Olympics and Other Sports" category. In addition to tapes which cover the Olympics and those which each review several sports, tapes in this section cover sports such as auto racing, billiards, bowling, skiing, and track and field, among others.

To further help you locate a tape, you'll find an index of titles, broken down by sport, at the end of the book. Tapes are listed alphabetically by the first word in the title. An exception to this is titles beginning with numerals—such as "1985 St. Louis Cardinals: Heck of a Year" or "10 Fundamentals of the Modern Golf Swing"—which are listed at the end of each index. Tapes which start with a player's name are listed by the player's last name; "Larry Bird: Basketball Legend" can be found under Bird. Lastly, titles beginning with words such as "The" and "A" can be found listed under the title's second word.

How Can You Get 'em?

Unless you're just insatiably curious about the world of video, you probably bought this book with the idea of adding to your collection. And so far, I haven't told you. All right, drum roll! Here's how.

The simplest way to go is to walk into your local video emporium and pick something off the shelf. Video stores deal mainly in movies. Sports videos are only a small percentage of their business. The bigger the store, the more likely you'll see something to your liking, but your choices will definitely be limited. And, if you're in a situation like I am, forget it. The biggest store in my neighborhood, Mom and Pop's Video and Live Bait Emporium, has had the same single sports video in stock for years.

The biggest chain of video stores in the country is Blockbuster Video, with about 2,000 franchises at last count. Although individual stores may vary, the folks at Blockbuster's home base say that you can expect to see a couple of dozen sports titles—mostly the very latest stuff—on their shelves, but that they will usually special order for you.

Blockbuster should be of special interest to baseball fans, in that they recently signed an exclusive contract with Major League Baseball. That doesn't mean that you won't be able to get a World Series video somewhere else; but if you order it from someone else, that store may have to order from Blockbuster to get it.

The second simplest way to get a video is to ask your local store to order it for you. They'll contact their distributor (who often will contact someone else), and eventually, you'll get your tape. However, you'll usually pay a price. All those middle men will expect to get something for their efforts. Of course, if money is no object, this is the easiest way to go. Shucks, you can even pay somebody else to watch the thing for you.

One problem. Tapes are like books in that sooner or later they "go out of print," that is, disappear from a distributor's shelves. When that happens, finding the one you're seeking becomes a treasure hunt. If word comes back to your local store that the tape you want is no longer available, all you can do is start writing or calling the mail order houses or place a classified in *Video Shopper*.

I wondered whether I should list those videos that may well no longer be in release by the time you pick up this book. I didn't have any sleepless nights over it, but I missed one afternoon nap. I finally decided to list everything on the theory that even something a year or two out of release may be gathering dust on someone's shelf (and might even be discounted). If you never heard of it, you can't order it.

Mail Order

The broadest selection of videos available to most people is through one of the big mail order houses. They'll be glad to send you a catalog, or you can call and make arrangements over the phone.

Sports Books, etc.
5224 Port Royal Road
Springfield, VA 22151
(703) 321-8660

Paul Haas, the owner, says he has the largest selection of sports books in the U.S. I'm interested in the "etc."—namely sports videos—and I'll have to say he probably leads that league, too. Sports Books, etc. is particularly strong in the major sports—baseball, football, and basketball—but it also has at least a few videos on just about any sport you can think of. At least 70 percent of the videos I have named here are listed in SBe's catalog, which is certainly worth perusing.

The listed prices seem to be in line with the suggested retail prices I found elsewhere, but you do have to pay for postage and handling. Contact the vendor for the latest rates.

Virginia residents are also subject to a 4.5 percent sales tax on the videos. MasterCard and VISA accepted. And there is a minimum charge of $20, so, if you only want a single $6.95 video, you're out of luck.

T.V. Sports Video Inc.
P.O. Box 1003
Darien, CT 06820
also (800) 729-0360 for VISA or Master Card orders.
or Fax to (203) 655-1486

T.V. Sports has a few baseball, basketball, and football videos in its catalog, but mostly they deal in boxing and wrestling. Most of their prices seem about $5 higher than the prices I saw elsewhere, but this was the only place I saw several boxing titles listed. They also had a line of baseball and football neckties that are absolutely god-awful. I hid the catalog from all my friends and relatives for fear they might consider ordering one as my next birthday gift.

There's a 6 percent sales tax for Connecticut residents. Allow six weeks for delivery, but if you just can't wait, you can get second day UPS shipping for an additional $7.50 per item.

Doak Ewing / RARE SPORTSFILMS
1126 Tennyson Lane
Naperville, IL 60540
(708) 527-8890 "anytime"

Doak Ewing's passion for old sports films led him into this decidedly specialized field. His product is unique. If you write to him, he'll send you a whole page of information on each of his tapes. As mentioned above, you can purchase his tapes from Doak himself or most of them from mail order houses.

One of the things that Ewing is most proud of is the restored color. That's *restored*, not *added*. Ewing's restorations are done with a computerized process by the same company that colorizes old movies. Colorization is a dirty word in some circles. You may have seen some of those old Bogart flicks where his skin is khaki. Fortunately, there's a world of difference between adding color to a black-and-white film that should have been left that way and bringing back color that has faded on an original colored film.

Add $2.50 for first class shipping. Beta format is also available for most tapes.

Golf Videos USA
675 East Big Beaver Road, Suite 209
Troy, MI 48083
(800) 362-GOLF, (313) 528-8071

GV USA bills itself as "America's largest golf video source," and with some 300 titles available, they'll get no argument here. Fortunately, they offer a free catalog.

Sports Direct, Inc.
PO Box 342
South Windsor, CT 06704
(800) 245-3011

Sports Direct advertises a string of World Series videos as well as quite a few baseball audios. It's unclear how Blockbuster's contract with Major League Baseball will affect Sports Direct's offerings, so call first.

K5 International
21 Murray Park Road
Winnipeg, Manitoba
R3J 3S2
800-665-5021

A Canadian-based company, K5 offers a limited number of sports videos on a wide variety of subjects.

PRODUCER/DISTRIBUTOR CODES

ATL	Atlas
BFS	BFS, Limited
BFV	Best Film & Video
CBS	CBS/Fox
CMV	Sony Music Video
CY	Celebrity
EM	Embassy
ESP	ESPN Home Video
FH	Fox Hills Video
FRI	Fries Home Video
FRV	Front Row Video
GD	Golf Digest
HBO	HBO Video
HNP	Heritage Now Production Group
HPG	HPG Home Video, Inc.
IVE	International Video Entertainment, Inc.
JCI	JCI Video
J2	J2 Communications
KOD	Kodak Video Programs
LAP	LA Production Group
MA	Master Vision
MLB/BLK	Major League Baseball/Blockbuster Entertainment
MOR	Morris Video
NAC	NAC Home
NBP	Norman Baer Productions, Inc.
PAR	Paramount
PHI	Premiere Home Video
PHO	Phoenix Communications Group
QV	Quality
RAY	Raycom Sports and Jefferson Pilot Teleproductions
RSF	Rare Sportsfilms
RVI	Rotfeld Video, Inc.
SBL	Scholarship Basketball Ltd.
SCU	Sports Clinic USA
SIM	Simitar Entertainment, Inc.
SYB	Sybervision
TW	Time Warner Home Video
VES	Vestron Video
VID	VidAmerica
VR	Video Reel
WES	Westcom Productions
WKV	Wood Knapp Video
WV	Worldvision Home Video

BASEBALL

Baseball is the national pastime, and, as such, the subject of every kind of sports video. You got a favorite team? You can probably find a video on its whole history or a highlight video of any of the last few seasons. Want to rerun a World Series or All Star Game? Unless you're on Medicare, most of the ones you remember are available. Your favorite player? If he was a superstar, especially a recent superstar, his life and diamond accomplishments have no doubt been immortalized on tape. At the other end of the spectrum are baseball blooper videos. Some superstars show up on those, too. Instructionals are available for just about anything legal you might want to do, but there are more for baseball than any other team sport. Likewise, Hollywood has always found baseball more compatible with the big screen than any sport except possibly boxing; some fine movies (and a lot of bad ones) are out there.

Team Highlight Videos

The purpose of marketing a team's highlight video is not to inform you about last year's team. Journalism and history are left to other media. If you want a true picture of what happened to your team last season, read newspapers and magazines, listen to the sports talk shows on radio and watch those on TV, talk with your fellow fans, and—above all—think. Don't watch a highlight video to learn; watch to enjoy. The reason your team is hawking that highlight video isn't to educate you, it's to get your mouth watering for next season. Hopefully, you'll salivate enough to rush right out and buy a season ticket.

The objective is the same whether that season ticket is for baseball, basketball, football, hockey, or professional tag-team checkers. And if baseball seems a bit more overt in selling you tomorrow by reliving yesterday, it may only be because of the nature of the sport or the length of the season. Perhaps they're all equally transparent. But the perception persists that baseball does more propagandizing than the others in its highlight videos.

That's not to say that your team's video is full of inaccuracies, fabrications, or lies. No seasoned propagandist is going to try to pass off a strikeout as a home run. He simply looks for the best face and then wears it to the party. It's a matter of spin.

No matter how good or bad a team might be, its 162-game season is sure to provide enough home runs and sparkling catches to fill a reel of tape. Unfortunately, one homer looks pretty much like another and the best fielding play in a month may not have any real bearing on the outcome of a particular game. For every game-saving catch, you'll see a half dozen in the late innings of 10–2 losses.

The main problem, of course, is that in any given season through 1991, there was one final winner and 25 losers. The latter number will increase by two with the addition of the Colorado Rockies and Miami Marlins to the National League. When the scriptwriter sits down to glorify the local nine, the odds are overwhelming that his ending is already written in failure. So, how does he get around it?

Let's take a quick peek at the way three highlight videos handled the National League race in 1990:

Wire to Wire takes you through the Cincinnati Reds' season to a pennant, a playoff victory, and a World Series triumph. What more could you ask?

Well, perhaps a little more drama. The Reds won their flag, as the title says, wire-to-wire, taking over first place with an opening day, 11-inning win at Houston and staying out in front of their Western Division foes for every single moment of the season. Larry Keith, doing the narration, stresses the hard work that went into making the victory possible, but that's a hard sell in a season when a team is never in serious jeopardy. The video attempts to build a little suspense when the Cincinnati lead drops to four-and-a-half games in August—heavens!—but soon the Reds are back on the track. The drama quotient increases slightly in the League Championship Series victory over Pittsburgh, and several outstanding defensive plays are worth seeing again. But once the Reds are in front three games to one, the result is foregone. The World Series is disposed of in a quick four straight.

Typical of highlight videos, interspersed throughout are short player profiles that provide nearly as much insight as you can find on the back of a bubble gum card. The Reds' video also lets you see owner Marge Schott giving her opinions again and again... and again. Is it only my imagination that she shows up more often than Manager Lou Piniella?

In *"No Doubt About It!"* the Pittsburgh Pirates' 1990 video, the drama quotient is up marginally because the Pirates had a marginally closer race than the Reds, even though they won more regular season games than Cincinnati. The video's title comes from Pirate announcer Lanny Frattare's signature sign-off to every Pittsburgh victory, "And there was no-oo-oo-oo doubt about it!" Well, actually there *was* some doubt about it during the season, but not enough to keep you on the edge of your seat in video review.

The player profiles break no new ground, but they do seem to give slightly better outlines of some of the Pirate personalities than we get of the Reds in *Wire to Wire*. Several Pirates enjoyed career years in 1990—Barry Bonds won the N.L. MVP, Bobby Bonilla finished second, and Doug Drabek won the Cy Young—and they are interviewed heavily during the course of the video. The Pirate ownership never appears, leaving more time to talk with the players.

In the end, of course, the Pirates lose the LCS to the Reds. But after relating this sad conclusion to the season, Frattare is quickly upbeat again in the coda about how Pirate fans "should feel proud" and all the rest of the stuff that hints that 1991 will be even better. Ironically, the 1991 season turned out to be a repeat, with Pittsburgh again romping through its division only to lose in the LCS.

One of the more interesting contrasts between *Wire to Wire* and *"No Doubt About It!"* is to be found in their reporting of a play that appears in both. Paul O'Neill's double in the second game of the LCS led to a Cincinnati win and was one of the turning points of the series. In the Reds' video, it is reported simply as a two-base hit; in the Pirates' video, Barry Bonds' misplaying of the ball is noted. The difference is a bow to fans' attitudes: for a Reds' fan, victory came about through Cincy efforts; for a Pirates' fan, the loss was the result of a rare Pittsburgh mistake. Both sides can feel they *deserved* to win.

The writer of *The 1990 New York Mets* couldn't say that. The

Mets went into 1990 *expected* to win and instead fell flat on their faces. For a while it looked like they were hellbound for last place. Then, after climbing back into the race, they lost key games to the Pirates and others to fall by the wayside again. So how did the writer handle what was undeniably a very disappointing season?

Well, he couldn't close his eyes and make it go away of course, but you'll find a lot of the emphasis is placed on how much better the Mets played after Buddy Harrelson replaced Davy Johnson as manager. The impression intended is that things would go right in the world in '91 with Buddy at the helm all the way. Little did the writer know that Harrelson would become the New York media's favorite whipping boy in 1991 and be replaced before that season's end.

Lacking any kind of a happy ending to this tale, *The 1990 New York Mets* focuses more on individual accomplishment. Fortunately, the Mets as individuals finished with more to show for 1990 than the Mets as a team. Mets fans could re-enjoy the hitting of Dave Magadan, Darryl Strawberry, and Howard Johnson, the pitching of Doc Gooden, Frank Viola, David Cone, and John Franco, and the all-around play of surprising Daryl Boston. After all, the real message is "wait 'til next year, but don't wait to buy your season tickets."

A statistician might find a correlation between the poor performance of a team and the amount of optimism present in its video. The worse things were last season, the better they'll be next year. Had the horrible St. Louis Browns of the 1930s had videos, they might have smiled the country out of the Depression.

Just about any team highlight video is going to be well-shot and relentlessly upbeat. Viewer preference depends almost completely on the viewer's team loyalties. A Reds fan may find a whole winter of content watching *Wire to Wire* over and over, but it's hard to imagine him sitting all the way through *Beating the Odds* (the 1990 Boston Red Sox) one time without wondering if he might be missing a great tractor pull on ESPN.

As an historical addendum, it would seem the writers of team highlight videos probably hate player free agency almost as much as some of the owners. It must be murder to put a wait-till-next-year spin on your story when you don't know whether some of your most important players will return. Imagine being puffed with pride (and trying to sell Pirates season tickets) with Bobby Bonilla's bashes when the next year he'll be bashing for the hated Mets.

History Videos

Team histories like *St. Louis Cardinals: The Movie* lack the immediacy of last year's highlight video, but in the long run may have more staying power for the viewer. Unless a season was particularly memorable, you may not have much inclination to watch it again five years from now, but a whole history of your favorite team can be fun again and again. Of course, the early years are culled almost exclusively from still photos and short, blink-and-you-missed-it snippets of film, but they have their charm too.

Reviews of seasons—like *This Year in Baseball 1988* or any other year—are okay, even if it always seems like interesting stories get short shrift and stories you could care less about go on forever. Obviously the makers are stuck with limited time, a variety of material, and making tough choices.

Special event videos—the All Star Game or World Series—sell baseball in general but are not tied to any particular team's ticket sales. The result is a better balanced, more journalistic approach. You still won't learn anything you didn't hear or see when the event took place (assuming you were paying attention). Videos are not investigative reporting; they just rehash the same stories that were on TV when the game or Series was news. But you may be reminded of some things you'd forgotten.

The 1990 World Series was a triumph for the Reds, but when The *1990 World Series* video leads off with five minutes on Oakland's pennant victory and five more on the Cincinnati's, you begin wondering how a Series that went only four games can fill up the 50 minutes of tape that are left.

It's done partly with so many crowd-reaction shots that a few of those fans begin to seem like family members. It's a matter of taste, but probably most viewers would rather watch a routine groundout than some fat guy in the stands biting his lip.

Fortunately, the '90 Series had a couple of good human interest stories. One of the best was Series MVP Jose Rijo's relationship with his father-in-law, Hall of Fame pitcher Juan Marichal, who ironically worked for Oakland. We get interviews with Rijo, his wife Rosie, and his father-in-law.

Then there's the story of Tom Browning's baby. Browning, you'll recall, left the ballpark early during Game Two when his wife was about to deliver. The game went into extra innings and the Reds began to run low on pitchers. Word was sent out over the radio and then the TV for Browning to get back to the park P.D.Q. Happily, the Reds won before Tom was forced to leave the hospital.

The video has the usual jock interviews. The big word was "confidence," as in "my homer/single/catch gave us confidence." Apparently it's no longer sufficient to make the play; one must also rejuvenate a team's psyche.

Running throughout the Series is the basic story of an underdog against the favorite, probably best expressed in the newspaper headline: REDS STUN GOLIATH. That's a scenario that never grows old, and there are enough shots of Reds doing the right thing and Athletics doing the wrong to convince us that David deserved his win.

Pat O'Brien's narration is okay except when he's forcing a God-awful pun, like the Reds were "Sabo-taging" the A's, or calling a play involving catcher Jamie Quirk a "quirk of strategy." Generous use is made of cuts from Jack Buck and Tim McCarver's telecast descriptions.

If you're at all a history buff, two videos that should not be missed are *The Boys of Summer*, based on Roger Kahn's best-seller about the Brooklyn Dodgers of the 1950s, and *The Glory of Their Times*, based on Lawrence Ritter's classic book of interviews with stars of the 1900–20 period of baseball. Both are textbook examples of how to make a very good video from a great book.

In 1994, Ken Burns' take on baseball is due on PBS and will no doubt be available on video the next day. If he can come anywhere near his work on the Civil War, his will be *the* history video to buy.

Stars and Superstars

Team game or no, individual personalities make fun baseball viewing. Well, most of the time anyway. Let's face it, some players are more interesting on the diamond than off. Still, if a star is big enough, somebody will try to put him on tape. Everybody from Hank Aaron to Ted Williams has a video life, telling you why he's more admirable than Mother Theresa. These "biographies" are pure vanilla, naturally. You'll never see anything in any of them that would excite the *National Enquirer*. On the other hand, what's wrong with having a few heroes?

Nolan Ryan: Feel the Heat has some of the best and worst traits of the genre. It gets off to an embarrassing start by adapting the theme from the old *Rawhide* TV show ("Rollin', rollin', rollin'—raw-HIDE!") to "Nolan, Nolan, Nolan—ry-YAN!" This gets worse with such lyrics as "A true-life Texas hero / Stackin' up the zeros." Another weakness is Ryan's own laid-back delivery in interviews; the man discusses a no-hitter as if he were explaining the purchase of a pair of shoes. The no-hitters are another problem; only the first five are mentioned. Finally, you realize the video was made after his 5,000th strikeout but before no-hitters six and seven. Before his 300th win, too. You may feel a little short-changed in not hearing how he felt about those two milestones.

And why was there no allusion made to Ryan's spartan daily workouts—the most important factor in keeping his magnificent arm young all these years? Of all his admirable qualities, and there are quite a few, it would seem his willingness to punish his body for his art would rank right up there. Yet, we're left with the impression that his longevity just happened through luck.

The major weakness in the video, of course, is that Ryan does one thing better than anyone else—throw a baseball past a hitter. So, we are left with seemingly endless shots of third strikes. With an Aaron or a Williams, the home runs can be leavened out with an occasional catch or stolen base or some*thing* besides Strike Three! Do we have to see all 5,000? But, to give the makers credit, they try to spice things up. One segment in which a camera is attached to a catcher's mask lets you watch that fastball roar in. If you ever wondered if you could hit it, this will convince you the answer is a resounding NO. Even better is a series of shots from behind the pitcher that superimposes a crackling flame on the strike zone as his pitch zooms over home plate. It's corny, but it actually gives you a feeling for how unhittable those hummers are.

Perhaps the most interesting part of all has Professor Robert Adair explaining exactly why Ryan's fastball is so hard to hit. It's not just the speed; it's what that speed makes the ball do! Ryan, in his flat Texas drawl, also shows how he throws his fastball, curve, and changeup, in a section that is perhaps of most interest to young pitchers.

Like every other sports video, the viewer's predilections to a large extent dictate how much or how little he'll like any biographical video. If you find Nolan Ryan an admirable human being with a remarkable skill, you'll forgive most of this video's weaknesses as I did.

Instructionals

In this section, I'm going to assume you're a proud papa. Instructionals are aimed primarily at fathers who want to help their sons (or daughters) play better baseball. Entertainment is not a main consideration, but, on the other hand, it doesn't hurt. After all, you want your kid to sit enthralled through this and not feel like he's going to the dentist. Seeing your youngster's favorite player on the screen can help keep his interest, but many fine instructionals have been made by players who were active before the child's time or by people he never heard of like college coaches.

Before you run out and buy an instructional for your son, make sure you know his attachment to the game. First, is his interest in baseball deep enough that he'll sit still and pay attention? Does he really want to improve his game or does he just want to have fun with the gang? If the latter, don't force him. "Shut up and watch!" can turn a kid off baseball and toward Barbie and Ken.

Another important thing to know is at what level your kid is playing now and at what level is he able to learn. If he's just a beginner or very young, the instructional you buy may go right over his head. On the other hand, it's possible that your kid is so advanced he'll be bored with a basic video. Not likely, but possible. If at all possible, preview any instructional before you shell out cash. You're buying it to do a specific job; take the time.

For example, a title like *Play Ball with Mickey Mantle, featuring Gary Carter and Tom Seaver* could set any papa's heart a-pounding. What those guys don't know about baseball isn't worth knowing. But, when you get it home, you'll find that it's made for more advanced baseballers of high school age or more. If your youngster is just at the beginning stage, he's going to be confused. Most of Mickey's tips won't do him much good for a couple of years. Although Mantle, Carter, and Seaver make reference to an earlier, more basic Mantle video, *Baseball Tips for All Ages*, you're going to wish that you had that one instead. In *Tips*, Mickey is assisted by Whitey Ford and—Holy Cow!—Phil Rizzuto. But the best part is that it starts with basics, like warming up, choosing proper equipment, and playing safely. Another useful feature is that the sleeve provides a chart which can be filled in from your VCR tape counter to make finding various tips easier. Once your youngster has mastered the basics and becomes more advanced in knowledge and ability, there's a lot of solid information in *Play Ball... Seaver,* in particular, appears to be a natural teacher.

One tip on choosing the proper video for your youngster's age level: look at any kids that appear in the video. They should be approximately the same age as yours.

Make sure you get the right video for the specific skills you and your boy want to improve. Okay, if you want to make him into a great pitcher, you're not likely to buy *The Art of Hitting .300*. But your choices are not that obvious. There are excellent instructionals on hitting, pitching, fielding, baserunning, warm-up exercises, and even coaching psychology. Then there are others that try to cover the whole range in one set. So, you should decide: is it his whole game that needs work or just part of it?

Once you have the video back home, you're only beginning. Don't just hand it to your kid and expect him to run off and watch it while you read your newspaper. Watch it with him! Especially the first time through. Seeing that you are interested will increase his interest. More important, there are bound to be concepts and terms that he won't understand. You should be there to push the pause button and enlighten him. You may find yourself trying to define a non-baseball word too, so having a dictionary near is a good idea. Don't be afraid to rerun important parts or to stop after a segment and discuss it.

Something else you should know if you buy several instructionals—there's more than one way to skin a cat and more than one way to hit a baseball. For example, the Charlie Lau school of hitting is different from the Ted Williams school. Don't let your kid be confused by two (or more) "right" ways to do something. Explain to him that there *are* different ways that work for different people and it's up to him to discover what works best for him.

Finally, after viewing the tape, why don't you go out in the backyard and work on some of the concepts. It'll be good for your kid and better for you.

Humorous Videos

Paul Haas, head of Sports Boooks, etc., one of the nation's largest mail order houses, says that humorous baseball tapes are among his best sellers. If you can get a laugh out of watching a third baseman bounce off a fence, outfielders misjudging fly balls, or players in the dugout wearing their hats funny, more power to you. Somehow, collections of these things never roll me in the aisles.

Why is it that a silly incident in a real game can set us guffawing for five minutes, while a similar incident in a blooper collection earns only a smile? Probably because of something every stand-up comedian knows. Catch 'em when they're not expecting it. Real baseball at the major league level is played with a grim earnestness. When something even moderately amusing happens, we're caught off guard. As a result the whatever-it-was seems a whole lot funnier than it really is. But, when they tell you ahead of time, "Hey, this is a scream!" every fiber of our being reacts: Oh yeah? MAKE me laugh!

Super Duper Baseball Bloopers, culled from the *This Week in Baseball* TV show, works hard at getting us to giggle, but with only occasional success. It has the usual misplays and dumb plays replete with clever sound effects. One section called "Field of Bad Dreams" even adds a perfectly awful Boris Karloff imitation. Another section wherein players relate incidents they thought were funny falls flatter than the path from first to second.

Although too much of its humor is forced, one incident in *Super Duper* broke me up. A pitch sails high over Willie Stargell's head, and he looks at the catcher with such an enigmatic expression that anyone with a funnybone can immediately think of a dozen punch lines. Fortunately, the scripters of the video left it to the viewer to pick one. THAT was funny!

Baseball Videos

Highlights: World Series

Most of the World Series and All Star Game Videos listed below are produced by Major League Baseball. All those from 1943 to 1983 are listed at 30 minutes, but that is more like a "ballpark" figure and many run considerably longer. Rare Sportsfilms, marked (RSF), also produces a number of World Series videos, usually piggybacked with some additional footage involving an interview.

75 YEARS OF WORLD SERIES MEMORIES. (Made in 1979). (MLB/BLK) BW/CO 30 min., $24.95.

20 YEARS OF WORLD SERIES THRILLS: *1938–1958*. (MLB/BLK) BW/CO 40 min., $24.95.

1943 WORLD SERIES, Yankees vs. Cardinals. (MLB/BLK) BW 30 min., $24.95.

1944 WORLD SERIES, Cardinals vs. Browns. (MLB/BLK) BW 30 min., $24.95.

1944 WORLD SERIES, Cardinals vs. Browns, with a 1953 interview of Bobo Holloman. (RSF) 29 min., $29.95.

1945 WORLD SERIES, Tigers vs. Cubs. (MLB/BLK) BW 30 min., $24.95.

1945 WORLD SERIES, Tigers vs. Cubs, plus 1939 Cubs in spring training, Catalina Island. (RSF) BW, 40 Min., $29.95.

1946 WORLD SERIES, Cardinals vs. Red Sox. (MLB/BLK) BW 30 min., $24.95.

1947 WORLD SERIES, Yankees vs. Dodgers. (MLB/BLK) BW 30 min., $24.95.

1948 WORLD SERIES, Indians vs. Braves. (MLB/BLK) BW 30 min., $24.95.

1948 WORLD SERIES, Indians vs. Braves, plus Feller-Hegan interview. (RSF) BW 70 min., $34.95.

1949 WORLD SERIES, Yankees vs. Dodgers. (MLB/BLK) BW 30 min., $24.95.

1949 WORLD SERIES, Yankees vs. Dodgers, plus regular-season footage of 1949 Cardinals and Yankees. (RSF) BW 55 min., $29.95.

1950 WORLD SERIES, Yankees vs. Phillies. (MLB/BLK) BW 30 min., $24.95.

1951 WORLD SERIES, Yankees vs. Giants. (MLB/BLK) BW 30 min., $24.95.

1951 WORLD SERIES, Yankees vs. Giants, plus interviews with Dark, Maglie, McDougald, Bauer. (RSF) 65 min., $34.95

1952 WORLD SERIES, Yankees vs. Dodgers. (MLB/BLK) BW 30 min., $24.95.

1953 WORLD SERIES, Yankees vs. Dodgers. (MLB/BLK) BW 30 min., $24.95.

1954 WORLD SERIES, Giants vs. Indians. (MLB/BLK) BW 30 min., $24.95.

1955 WORLD SERIES, Dodgers vs. Yankees. (MLB/BLK) BW 30 min., $24.95.

1955 WORLD SERIES, Dodgers vs. Yankees, plus Series preview. (RSF) BW 38 Min., $29.95.

1956 WORLD SERIES, Yankees vs. Dodgers. (MLB/BLK) BW 30 min., $24.95.

1956 WORLD SERIES, Yankees vs. Dodgers, plus Series preview. (RSF) BW 48 Min., $29.95.

1957 WORLD SERIES, Braves vs. Yankees. (MLB/BLK) BW 30 min., $24.95.

1957 WORLD SERIES, Braves vs. Yankees, plus Series preview. (RSF) BW 48 Min., $29.95.

1958 WORLD SERIES, Yankees vs. Braves; first Series to be filmed in color. (MLB/BLK) CO 30 min., $24.95.

1958 WORLD SERIES, Yankees vs. Braves. (RSF) CO 40 Min., $29.95.

1959 WORLD SERIES, Dodgers vs. White Sox. (MLB/BLK) CO 30 min., $24.95.

1959 WORLD SERIES, Dodgers vs. White Sox; Series in color, plus three other games from 1959 in BW. (RSF) 43 Min., $29.95.

1960 WORLD SERIES, Pirates vs. Yankees. (MLB/BLK) CO 30 min., $24.95.

1960 WORLD SERIES, Pirates vs. Yankees; in color, plus Elroy Face interview in BW. (RSF) 52 Min., $29.95.

1961 WORLD SERIES, Yankees vs. Reds. (MLB/BLK) CO 30 min., $24.95.

1961 WORLD SERIES, Yankees vs. Reds; in color, plus Roger Maris's 61st homer in BW. (RSF) 40 Min., $29.95.

1962 WORLD SERIES, Yankees vs. Giants. (MLB/BLK) CO 30 min., $24.95.

1962 WORLD SERIES, Yankees vs. Giants. (RSF) CO 39 Min., $29.95.

1963 WORLD SERIES, Dodgers vs. Yankees. (MLB/BLK) CO 30 min., $24.95.

1963 WORLD SERIES, Dodgers vs. Yankees. (MLB/BLK) (RSF) CO 37 Min., $29.95.

1964 WORLD SERIES, Cardinals vs. Yankees. (MLB/BLK) CO 30 min., $24.95.

1964 WORLD SERIES, Cardinals vs. Yankees. (RSF) CO 40 Min., $29.95.

1965 WORLD SERIES, Dodgers vs. Twins. (MLB/BLK) CO 30 min., $24.95.

1966 WORLD SERIES, Orioles vs. Dodgers. (MLB/BLK) CO 30 min., $24.95.

1966 WORLD SERIES, Orioles vs. Dodgers. (RSF) CO 39 Min., $29.95.

1967 WORLD SERIES, Cardinals vs. Red Sox. (MLB/BLK) CO 30 min., $24.95.

1967 WORLD SERIES, Cardinals vs. Red Sox. (RSF) CO 40 Min., $29.95.

1968 WORLD SERIES, Tigers vs. Cardinals. (MLB/BLK) CO 30 min., $24.95.

1969 WORLD SERIES, Mets vs. Orioles. (MLB/BLK) CO 30 min., $24.95.

1970 WORLD SERIES, Orioles vs. Reds. (MLB/BLK) CO 30 min., $24.95.

1971 WORLD SERIES, Pirates vs. Orioles. (MLB/BLK) CO 30 min., $24.95.

1972 WORLD SERIES, A's vs. Reds. (MLB/BLK) CO 30 min., $24.95.

1973 WORLD SERIES, A's vs. Dodgers. (MLB/BLK) CO 30 min., $24.95.

1974 WORLD SERIES, A's vs. Dodgers. (MLB/BLK) CO 30 min., $24.95.

1975 WORLD SERIES, Reds vs. Red Sox. (MLB/BLK) CO 30 min., $24.95.

1976 WORLD SERIES, Reds vs. Yankees. (MLB/BLK) CO 30 min., $24.95.

1977 WORLD SERIES, Yankees vs. Dodgers. (MLB/BLK) CO 30 min., $24.95.

1978 WORLD SERIES, Yankees vs. Dodgers. (MLB/BLK) CO 30 min., $24.95.

1979 WORLD SERIES, Pirates vs. Orioles. (MLB/BLK) CO 30 min., $24.95.

1980 WORLD SERIES, Phillies vs. Royals. (MLB/BLK) CO 30 min., $24.95.

1981 WORLD SERIES, Dodgers vs. Yankees. (MLB/BLK) CO 30 min., $24.95.

1982 WORLD SERIES, Cardinals vs. Brewers. (MLB/BLK) CO 30 min., $24.95.

1983 WORLD SERIES, Orioles vs. Phillies. (MLB/BLK) CO 30 min., $24.95.

1984 WORLD SERIES, Tigers vs. Padres. (MLB/BLK) CO 60 min., $24.95.

1985 WORLD SERIES, Royals vs. Cardinals. (MLB/BLK) CO 60 min., $24.95.

1986 WORLD SERIES, Mets vs. Red Sox. (MLB/BLK) CO 60 min., $24.95.

1987 WORLD SERIES, Twins vs. Cardinals. (MLB/BLK) CO 60 min., $24.95.

1988 WORLD SERIES, Dodgers vs. A's. (MLB/BLK) CO 60 min., $24.95.

1989 WORLD SERIES, A's vs. Giants. (MLB/BLK) CO 60 min., $24.95.

1990 WORLD SERIES, Reds vs. A's. (MLB/BLK) CO 60 min., $24.95.

1991 WORLD SERIES, Twins vs. Braves. (MLB/BLK) CO 60 min., $24.95.

M.V.P.: World Series Edition. (1979) (MLB/BLK) 60 min., $24.95.

OCTOBER SPOTLIGHT: *World Series Heroes: The Men Who Made It Happen.* (1980) (MLB/BLK) 30 min., $24.95.

ONCE IN A LIFETIME: *World Series Heroes.* (1981) (MLB/BLK) 30 min., $24.95.

GREAT WORLD SERIES HEROES: *The Men Who Made It Happen.* (1982) (MLB/BLK) 30 min., $24.95.

WORLD SERIES UNSUNG HEROES. (1983) (MLB/BLK) 30 min., $24.95.

Highlights: All Star Games

Maybe *you* are some kind of genius who can remember one All Star Game from another by the year. I can't. I'll add the location and hope that will help a little.

1948, 1952 & 1955 ALL STAR GAMES. All on one tape. (RSF) BW 35 Min., $29.95.

1956 ALL STAR GAME. At Griffith Stadium, Washington. In color, plus "Ruth to Mays" in BW. (RSF) 40 Min., $29.95.

1962 ALL STAR GAME. July 10, D.C. Stadium, Washington. (MLB/BLK) 30 min., $24.95

1962 ALL STAR GAME. July 30, Wrigley Field, Chicago. (MLB/BLK) 30 min., $24.95.

1965 ALL STAR GAME. July 13, Metropolitan Stadium, Minnesota. (MLB/BLK) 30 min., $24.95

1966 ALL STAR GAME. July 12, Busch Stadium, St. Louis. 10 innings. (MLB/BLK) 30 min., $24.95.

1967 ALL STAR GAME. July 11, Anaheim Stadium, California. 15 innings. (MLB/BLK) 30 min., $24.95.

1970 ALL STAR GAME. July 14, Riverfront stadium, Cincinnati. 12 innings. (MLB/BLK) 30 min., $24.95.

1971 ALL STAR GAME. July 13, Briggs Stadium, Detroit. (MLB/BLK) 30 min., $24.95.

1972 ALL STAR GAME. July 25, Fulton County Stadium, Atlanta. 10 innings. (MLB/BLK) 30 min., $24.95.

1973 ALL STAR GAME. July 24, Royals Stadium, Kansas City. (MLB/BLK) 30 min., $24.95.

1974 ALL STAR GAME. July 23, Three Rivers Stadium, Pittsburgh. (MLB/BLK) 30 min., $24.95.

1975 ALL STAR GAME. July 15, County Stadium, Milwaukee. (MLB/BLK) 30 min., $24.95.

1976 ALL STAR GAME. July 13, Veterans Stadium, Philadelphia. (MLB/BLK) 30 min., $24.95.

1977 ALL STAR GAME. July 19, Yankee Stadium, New York. (MLB/BLK) 30 min., $24.95.

1978 ALL STAR GAME. July 11, San Diego Stadium, San Diego. (MLB/BLK) 30 min., $24.95.

1979 ALL STAR GAME. July 17, Kingdome, Seattle. (MLB/BLK) 30 min., $24.95.

1980 ALL STAR GAME. July 8, Dodger Stadium, Los Angeles. (MLB/BLK) 30 min., $24.95.

1981 ALL STAR GAME. August 9, Municipal Stadium, Cleveland. (MLB/BLK) 30 min., $24.95.

1982 ALL STAR GAME. July 13, Montreal Stadium, Montreal. (MLB/BLK) 30 min., $24.95.

1983 ALL STAR GAME. July 6, Comiskey Park, Chicago. (MLB/BLK) 30 min., $24.95.

1984 ALL STAR GAME. July 10, Candlestick Park, San Francisco. (MLB/BLK) 30 min., $24.95.

1985 ALL STAR GAME. July 16, Metrodome, Minnesota. (MLB/BLK) 30 min., $24.95.

1986 ALL STAR GAME. July 15, Astrodome, Houston. (MLB/BLK) 30 min., $24.95.

1987 ALL STAR GAME. July 14, Oakland-Alameda Stadium, Oakland. (MLB/BLK) 30 min., $24.95.

1988 ALL STAR GAME. July 12, Riverfront Stadium, Cincinnati. (MLB/BLK) 30 min., $19.95.

Highlights: Baseball History, Seasons

Baseball's long history has so many great moments that no one tape can cover them all. Fortunately, there's no need to stop at one video. Don't overlook *The Glory of Their Times.*

BASEBALL'S GREATEST MOMENTS. The top 20 events in the game's history. 60 min., $19.95.

BASEBALL'S RECORD BREAKERS. Narrated by Warner Fusselle. 45 min., $14.95.

THE GLORY OF THEIR TIMES. Based on Lawrence Ritter's classic, this film by award-winning Bud Greenspan looks at 1896–1916 baseball heroes. (VID, 1987) 60 min., $14.98.

HISTORY OF BASEBALL: *Greatest Moments from Baseball's Past.* From the games's early history to today's teams. (MLB/BLK, 1987) 120 min., $29.95.

HISTORY OF GREAT BLACK BASEBALL PLAYERS. Hosted by Ernie Banks; highlights from the old Negro Leagues through the present. (FRI, 1991) 45 min., $19.95.

THIS WEEK IN BASEBALL'S GREATEST PLAYS. The best from T.W.I.B. over the years; narrated by Mel Allen. 45 min., $19.95.

GOLDEN DECADE OF BASEBALL 1947–1957. Narrated by Brent Musburger. The best of times in New York baseball; DiMaggio, Mays, Mantle, Snider, and so many others! (NBP, 1991)

   PART 1. (1990) 60 min., $14.95.

   PART 2. (1990) 60 min., $14.95.

1938 AMERICAN LEAGUE FILM: *The First Century of Baseball.* $29.95.

BASEBALL IN THE NEWS. (ATL, 1984).

   VOL. I (1951–55). Old newsreels. 60 min., $29.95.

   VOL. II (1956–60). Old newsreels. 60 min., $29.95.

   VOL. III (1961–67). Old newsreels. 60 min., $29.95.

1951 BASEBALL NEWS. Old TV baseball news. (RSF) 37 min., $29.95.

1955 WASHINGTON SENATORS: *Story of the Washington Nats.* A promo film that includes footage of Griffith Stadium, plus 1959 interview with Harmon Killebrew; narrated by Bob Wolff and Arch McDonald. (RSF) CO 35 min., $29.95.

1955–1956 BASEBALL NEWS. Old TV Baseball News as seen weekly in the '50s. (RSF) BW 84 min., $36.95.

BASEBALL NEWS HIGHLIGHTS 1959. (MLB/BLK) 71 min. $29.95.

BASEBALL 1968. (MLB/BLK) 30 min., $24.95.

BASEBALL 1969. 30 min., $24.95.

1970 NATIONAL LEAGUE. (MLB/BLK) 30 min., $24.95.

1970 AMERICAN LEAGUE. (MLB/BLK) 30 min., $24.95.

1975 NATIONAL LEAGUE. (MLB/BLK) 30 min., $24.95.

1975 AMERICAN LEAGUE. (MLB/BLK) 30 min., $24.95.

BASEBALL IN THE '70s. Narrated by Mel Allen. (MLB/BLK) 48 min., $19.95.

BASEBALL IN THE '80s. Narrated by Jack Brickhouse. (MLB/BLK) 65 min., $19.95.

DECADE OF TRANSITION: THE '70s. (MLB/BLK) 30 min., $24.95.

1986 THIS YEAR IN BASEBALL. (MLB/BLK) 30 min., $14.95.

YEAR IN BASEBALL 1988. (MLB/BLK) 60 min., $19.95.

THIS WEEK IN BASEBALL 1990. The greatest moments and plays of the year, narrated by Mel Allen. (MLB/BLK) 60 min., $19.95.

Team Histories, Seasons

More Yankees videos are available than for any other team. For those generations who grew up when the Yankees were baseball's dominant team, this is a bonanza. If you applauded them before, you can do it again. If you were among the legions of Yankee-haters, you can suffer anew.

Atlanta Braves
(also Boston, Milwaukee)

1947 BOSTON BRAVES: *The Braves Family.* The first baseball film ever made in color! (RSF) CO 40 Min., $29.95.

1954 MILWAUKEE BRAVES: *Home of the Braves.* Hank Aaron's rookie season. (RSF) CO 29 Min., $29.95.

1955 MILWAUKEE BRAVES: *Baseball's Main Street.* (RSF) CO 29 Min., $29.95.

1956 MILWAUKEE BRAVES: *Bravesland, U.S.A.* (RSF) CO 29 Min., $29.95.

1957 MILWAUKEE BRAVES: *Hail to the Braves!* World Championship season. (RSF) CO 29 Min., $29.95.

1959 MILWAUKEE BRAVES: *Fighting Braves of '59.* Includes NL playoff game vs. Dodgers. (RSF) 28 min., $29.95.

1960 MILWAUKEE BRAVES: *The Best of Baseball.* Includes footage of first-ever Milwaukee old-timers game. (RSF) 29 min., $29.95.

Baltimore Orioles

1962 BALTIMORE ORIOLES: *The Baltimore Orioles in Action.* Narrated by Chuck Thompson. (RSF) CO 27 min., $29.95.

1966 BALTIMORE ORIOLES. Highlights; narrated by Chuck Thompson. (MLB/BLK) $22.95.

1969 BALTIMORE ORIOLES. Highlight video., (MLB/BLK) $22.95.

1979–81 BALTIMORE ORIOLES. Highlights from three seasons. (MLB/BLK) $22.95.

1982 BALTIMORE ORIOLES. Cal Ripken, Jr.'s rookie year. (MLB/BLK) $22.95.

1983 BALTIMORE ORIOLES. Season highlights. (MLB/BLK) $22.95.

1984 BALTIMORE ORIOLES. Highlights of the season. (MLB/BLK) $22.95.

1989 BALTIMORE ORIOLES: Why Not? Highlights. (MLB/BLK) $24.95.

Boston Red Sox

FENWAY: *75 Years of Red Sox Baseball.* Life and times at one of baseball's greatest stadiums. (MLB/BLK,1987) 60 min., $19.95.

1954 BOSTON RED SOX: *Baseball in Boston.* First Red Sox promotional film; narrated by Curt Gowdy. (RSF) CO 22 Min., $29.95.

1955 BOSTON RED SOX: *The Red Sox at Home.* Includes Hall of Fame Day at Fenway and Hall of Fame game at Cooperstown vs. Braves; narrated by Curt Gowdy. (RSF) CO 21 Min., $29.95.

1956 BOSTON RED SOX: *Pride of New England.* Narrated by Curt Gowdy. (RSF) CO 21 Min., $29.95.

1957 BOSTON RED SOX: *Play Ball with the Red Sox.* Plus Frank Malzone interview in BW; narrated by Curt Gowdy. (RSF) CO 25 Min., $29.95.

1986 BOSTON RED SOX: *1986 A.L. Champions.* Double Header: Includes Team Highlights and This Year in Baseball. (MLB/BLK) 60 min., $19.95.

1988 BOSTON RED SOX: *Morgan's Magic.* (MLB/BLK) 60 min., $19.95.

1990 BOSTON RED SOX: *Beating the Odds*. Narrated by Curt Gowdy. (MLB/BLK) 50 min., $14.95.

California Angels

1962 LOS ANGELES ANGELS: *Angels '62*. Narrated by Don Wells. (RSF) CO 24 min., $29.95.

Chicago Cubs

CHICAGO AND THE CUBS: *A Lifelong Love Affair*. History of the Cubbies from Anson to Sandberg. (MLB/BLK, 1987) 60 min., $19.95.

CHICAGO'S GRAND STANDS: *Chicago's Classic Ballparks Then and Now*. Highlights at Wrigley and old Comiskey. 30 min., $14.95.

1969 CUBS VS. PHILLIES COMPLETE GAME. Every pitch of a single game at Wrigley Field. (RSF) BW 140 min. (2 tapes), $59.95.

1989 CHICAGO CUBS: *Boys of Zimmer*. Cubs win the N.L. East. (MLB/BLK) 60 min., $19.95.

Chicago White Sox

CHICAGO WHITE SOX: *A Visual History*. (MLB/BLK) 60 min., $19.95.

CHICAGO'S GRAND STANDS: *Chicago's Classic Ballparks Then and Now*. Highlights at Wrigley and old Comiskey. 30 min., $14.95.

Cincinnati Reds

REDS: *The Official History of the Cincinnati Reds*. (MLB/BLK) 60 min., $19.95.

1990 CINCINNATI REDS: *Wire to Wire*. (1990) (MLB/BLK) 60 min., $14.95.

Cleveland Indians

1986 CLEVELAND INDIANS: *Indian Summer.* Double Header: Includes Team Highlights and This Year in Baseball. (1986) (MLB/BLK) 60 min., $19.95.

Detroit Tigers

DETROIT TIGERS: *The Movie* (History of the Tigers). (MLB/BLK) 86 min., $29.95.

1958 DETROIT TIGERS: *Tigertown U.S.A.* and *1962: Baseball for Little Leaguers.* 1958 spring training and regular season action, plus an instructional film with the 1962 Tigers in spring training. (RSF) 51 min., $29.95.

Houston Astros

SILVER ODYSSEY: *25 Years of Houston Astros Baseball.* (MLB/BLK) 60 min., $19.95.

Kansas City Royals

1985 KANSAS CITY ROYALS: *Thrill of it All.* (MLB/BLK) 23 min., $19.95.

Los Angeles Dodgers
(also Brooklyn)

THE BOYS OF SUMMER. Based on Roger Kahn's best-seller; interviews with the great Dodgers players of the 1950s. (VID) 90 min., $14.98.

DODGER STADIUM: *25th Anniversary.* Great moments at Chavez Ravine. (MLB/BLK, 1987) 70 min., $19.95.

100 YEARS: *A Visual History of the Dodgers.* Dodger history from

1890–1990; hosted by Vin Scully. (J2) 76 min., $19.95.

1988 LOS ANGELES DODGERS: *Through the Eyes of a Winner.* (MLB/BLK) 60 min., $19.98.

Milwaukee Brewers

1986 MILWAUKEE BREWERS: *Family Album.* Double Header: Includes Team Highlights and This Year in Baseball. (MLB/BLK) 60 min., $19.95.

Minnesota Twins

THEN & NOW: *The Minnesota Twins' Silver Anniversary* (1961–1985). 27 min., $19.95.

1961 MINNESOTA TWINS. From spring training on, the Twins' inaugural season at the old Met. (RSF) CO 27 min., $29.95.

1986 MINNESOTA TWINS: *Power & Promise.* Double Header: Includes Team Highlights and This Year in Baseball. (MLB/BLK) 60 min., $19.95.

1987 MINNESOTA TWINS: *Twins Win* . (MLB/BLK) 60 min., $19.95.

New York Mets

AMAZIN' ERA: *New York Mets 25 Years.* (MLB/BLK, 1986) 71 min., $19.95.

1965 NEW YORK METS: *Expressway to the Big Leagues.* Includes Casey Stengel's uniform retirement ceremony. (RSF) CO 27 min., $29.95.

1967 NEW YORK METS: *Year of Change.* The year Seaver and Ryan were rookies. (MLB/BLK) 27 min., $29.95.

1986 NEW YORK METS: *A Year to Remember.* Double Header: Includes Team Highlights and This Year in Baseball. (MLB/BLK) 60 min., $19.95.

THE 1990 NEW YORK METS. (MLB/BLK) 40 min., $19.95.

New York Yankees

NEW YORK YANKEES: *The Movie.* History of Yankees from 1903–1986. (MLB/BLK) 100 min., $29.95.

10 GREATEST MOMENTS IN YANKEE HISTORY. Original footage and live play-by-play of Ruth, Gehrig, DiMaggio, Mantle, Maris, etc., hosted by Mel Allen. (KOD) 30 min., $19.95.

GAME OF THE WEEK: *Yankees vs. Washington Senators.* At Yankee Stadium, Sept. 22, 1954. Shows almost every play, including first triple play in Yankee Stadium history. (RSF) 27 min., $29.95.

1956 NEW YORK YANKEES: *Winning with the Yankees.* Narrated by Mel Allen. (RSF) CO 34 min., $29.95.

PINSTRIPE POWER: *Story of the 1961 Yankees.* Maris, Mantle, and Ford. (MLB/BLK, 1987) 49 min., $19.95.

1978 NEW YORK YANKEES: *Greatest Comeback Ever.* Bucky Dent and Phil Rizzuto with an insider's view of the '78 team's unbelievable season. (VID, 1987) 58 min., $14.98.

NEW YORK YANKEES GREAT GAME BROADCASTS. Each video contains the entire game, pitch-by-pitch. $49.95 each.

    RON GUIDRY'S 18 STRIKEOUT PERFORMANCE. June 17, 1978.

    YANKEES-RED SOX PLAYOFF GAME. Oct. 2, 1978.

    DAVE RIGHETTI'S NO-HIT GAME. July 4, 1983.

NEW YORK YANKEES SPECIAL THEME CASSETTES. $29.95 each.

    DYNASTY: *The New York Yankees.* History through 1981.

    50 YEARS OF YANKEE ALL STARS. Individual Yankee All Stars.

NEW ERA. A look at Yankee "eras" of play. (1984)

1978: *It Don't Come Easy.* The 1978 comeback team; narrated by Bill White.

PLAY BALL WITH THE YANKEES. Instructional tape with the 1950 Yankees.

WINNING TRADITION. A look at the 1977 season and the Yankee tradition over the years.

YANKEE STADIUM: *Home of Heroes.* A look at the historic stadium and the stars who played there.

Oakland Athletics
(also Philadelphia, Kansas City)

1956 KANSAS CITY ATHLETICS: *The Kansas City A's in Action.* Includes footage of Connie Mack Stadium and other AL parks. (RSF) CO 24 Min., $29.95.

OAKLAND A's: *All Star Almanac.* A's in All Star Games. (MLB/BLK) 25 min., $19.95.

1988 OAKLAND A's: *A Bashing Success.* Includes Canseco's 40/40 season. (MLB/BLK) 60 min., $19.95.

1989 OAKLAND A's and SAN FRANCISCO GIANTS: *Champions by the Bay.* Includes regular season, playoffs, and earthquake-interrupted World Series for both teams. (MLB/BLK) 90 min., $19.95.

1990 OAKLAND ATHLETICS and SAN FRANCISCO GIANTS: *A Call to Arms.* Includes regular season for both teams, plus Oakland playoffs and World Series. (MLB/BLK) 90 min., $19.95.

Philadelphia Phillies

CENTENNIAL: *Over 100 Years of Philadelphia Phillies Baseball.* 60 min., (MLB/BLK) $19.95.

1980 PHILADELPHIA PHILLIES: *We Win!* The team that wouldn't die. (MLB/BLK) $29.95.

1986 PHILADELPHIA PHILLIES: *Headed for the Future.* Double Header: Includes Team Highlights and This Year in Baseball. (MLB/BLK) 60 min., $19.95.

1988 PHILLIES' HOME COMPANION, VOL. I: *The Game's Easy, Harry.* 60 min., (MLB/BLK) $19.95.

1989 PHILLIES' HOME COMPANION, VOL. II: *Not Necessarily Another Day at the Yard.* (MLB/BLK) 60 min., $19.95.

1990 PHILLIES' HOME COMPANION, VOL. III: *Gettin' Dirty.* (MLB/BLK) 60 min., $19.95.

Pittsburgh Pirates

BATTLIN' BUCS: *The First 100 Years of the Pittsburgh Pirates.* (MLB/BLK) 60 min., $19.95.

1960 PITTSBURGH PIRATES: *We Had 'em All the Way.* Championship season climaxed by Mazeroski's home run; narrated by Bob Prince. (MLB/BLK) 30 min., $29.95.

1990 PITTSBURGH PIRATES: *No Doubt About It.* NL East champs; Bonds wins MVP, Drabek wins Cy Young. (MLB/BLK) 60 min., $14.95.

St. Louis Cardinals

ST. LOUIS CARDINALS: *The movie. 109 Years of the Cardinals.* (MLB/BLK) 90 min., $29.95.

1957 CARDINAL TRADITION AND THE GAME NOBODY SAW. Fictional account of Cardinals rookie interwoven with actual accounts of Cardinal World Series triumphs; also teaching fundamentals in 1959 spring training. 54 min., $29.95.

1985 ST. LOUIS CARDINALS: *Heck of a Year.* (MLB/BLK) 45 min., $19.95.

1987 ST. LOUIS CARDINALS: *That's a Winner.* (MLB/BLK) 60 min., $24.95.

San Francisco Giants
(also New York)

GIANTS HISTORY: *The Tale of Two Cities.* (MLB/BLK) 60 min., $19.95.

1986 SAN FRANCISCO GIANTS: *You Gotta Like This Team.* Double Header: Includes Team Highlights and This Year in Baseball. (MLB/BLK) 60 min., $19.95.

1989 OAKLAND A's and SAN FRANCISCO GIANTS: *Champions by the Bay.* Includes regular season, playoffs, and earthquake-interrupted World Series for both teams. (MLB/BLK) 90 min., $19.95.

1990 OAKLAND ATHLETICS and SAN FRANCISCO GIANTS: *A Call to Arms.* Includes regular season for both teams, plus Oakland playoffs and World Series. (MLB/BLK) 90 min., $19.95.

Seattle Mariners

DIAMOND IN THE EMERALD CITY: *10 Years of Seattle Mariners Baseball.* (MLB/BLK) 59 min., $19.95.

Texas Rangers

1986 TEXAS RANGERS. Double Header: Includes Team Highlights and This Year in Baseball. (MLB/BLK) 60 min., $19.95.

Toronto Blue Jays

MISSION IMPOSSIBLE: *The First Decade—10 Years of Toronto Blue Jays Baseball.* (MLB/BLK) 60 min., $19.95.

1989 TORONTO BLUE JAYS: *Sky High—A.L. East Champions.* (MLB/BLK) 55 min., $19.95.

BASEBALL IN THE '80S. Narrated by Jack Brickhouse. (1989) (MLB/BLK) 65 min., $19.95.

1990 TORONTO BLUE JAYS: *Tradition of Success.* (MLB/BLK) 55 min., $19.95.

Individual Player Profiles

(alphabetically by player's last name)

The Video Baseball Cards series (formerly Greatest Sports Legends) offers 30-minute highlight/biographical tapes at a low price.

HANK AARON. Video Baseball Cards. (RVI) 30 min., $6.95.

JOHNNY BENCH. Video Baseball Card. (RVI) 30 min., $6.95.

YOGI BERRA. Video Baseball Card. (RVI) 30 min., $6.95.

ROBERTO CLEMENTE: *A Touch of Royalty.* 30 min., $24.95.

ANDRE DAWSON: *He's a Hero.* Highlights of Andre's 1987 MVP season. 60 min., $19.95.

JOE DiMAGGIO. Video Baseball Card. (RVI) 30 min., $6.95.

DON DRYSDALE. Video Baseball Card. (RVI) 30 min., $6.95.

WHITEY FORD. Video Baseball Card. (RVI) 30 min., $6.95.

STEVE GARVEY. Video Baseball Card. (RVI) 30 min., $6.95.

LOU GEHRIG. Video Baseball Card. (RVI) 30 min., $6.95.

REGGIE JACKSON. Video Baseball Card. (RVI) 30 min., $6.95.

REGGIE JACKSON: *Mr. October.* Interviews and highlights. 35 min., $14.95.

MICKEY MANTLE: *The American Dream Comes to Life.* $19.95.

MICKEY MANTLE. Video Baseball Card. (RVI) 30 min., $6.95.

MICKEY MANTLE/WILLIE MAYS. Double Headers. (RVI), 45 min., $14.95.

BILLY MARTIN: *The Man, the Myth, the Manager.* Highlights and interviews covering Billy Martin's career. $19.95.

WILLIE MAYS. Video Baseball Card. (RVI) 30 min., $6.95.

STAN MUSIAL. Video Baseball Card. (RVI) 30 min., $6.95.

LEGEND OF STAN THE MAN MUSIAL. Hosted by Jack Buck; includes rare action photos, anecdotes, tributes. $19.95.

JIM PALMER. Video Baseball Card. (RVI) 30 min., $6.95.

BROOKS ROBINSON. Greatest Sports Legends. (RVI) 30 min., $6.95.

FRANK ROBINSON. Greatest Sports Legends. (RVI) 30 min., $6.95.

JACKIE ROBINSON/ROBERTO CLEMENTE. Double Headers. (RVI) 45 min., $14.95.

PETE ROSE. Video Baseball Card. (RVI) 30 min., $6.95.

BABE RUTH. Video Baseball Card. (RVI) 30 min., $6.95.

BABE RUTH: *The Man, the Myth, the Legend.* The Bambino's career, his personal life, and his impact on the game and our culture; hosted by Mel Allen. (1991) 40 min., $19.95.

MEET BABE RUTH. Highlights of the Babe's career. 30 min., $14.95.

MY DAD, THE BABE: *The Babe Ruth Scrapbook.* Based on the book by the Babe's daughter. $19.95.

BABE RUTH/JOE DiMAGGIO. Double Headers. (RVI) 45 min., $14.95.

NOLAN RYAN: *Feel the Heat.* The strikeout king in action. (1992, HPG) 53 min., $19.95.

NOLAN RYAN. Video Baseball Card. (RVI) 30 min., $6.95.

MIKE SCHMIDT. Video Baseball Card. (RVI) 30 min., $6.95.

OZZIE (Smith): *The Movie.* Highlights of the career of Ozzie Smith, Baseball's Wizard of Oz. 45 min., $24.95.

DUKE SNIDER. Video Baseball Card. (RVI) 30 min., $6.95.

TED WILLIAMS/PETE ROSE. Double Headers. 45 min., $14.95.

TED WILLIAMS. Video Baseball Cards. (RVI) 30 min., $6.95.

See also several biographical films.

Miscellaneous

BALL TALK: *Baseball's Voices of Summer.* Larry King hosts Mel Allen, Red Barber, Jack Brickhouse, Jack Buck, Curt Gowdy & Ernie Harwell as they reminisce, along with rare archival footage. (J2) 50 min., $29.95.

BASEBALL CARD COLLECTOR. Insider's guide to buying, selling, trading and collecting, narrated by Mel Allen. 35 min., $19.95.

BASEBALL DREAM TEAM (American League). Profiles of some of the A.L.'s greatest players. 32 min., $14.95.

BASEBALL DREAM TEAM (National League). Profiles of some of the N.L.'s greatest players. 32 min., $14.95.

BASEBALL DYNASTIES: *The New York Yankees, Oakland A's and Cincinnati Reds.* A look at three of the game's greatest dynasties. 30 min. $9.99.

BASEBALL'S GREATEST HITS. Musical videos, including "Take Me Out to the Ballgame," "Talkin' Baseball." 30 min., $14.95.

BASEBALL LEGENDS. Highlights of baseball's greatest players. (RVI), 40 min., $6.95.

BASEBALL TIME CAPSULE: *A Journey Through the Barry Halper Collection.* Hosted by Billy Martin, with Mantle, Mattingly & Ryan. 49 min., $14.95.

BASEBALL RIVALRIES: *The Yankees and the Dodgers.* Highlights of the eleven World Series meetings between the two. 30 min. $9.99.

50 YEARS OF LITTLE LEAGUE BASEBALL. The history of Little League; highlights and interviews. (CBS, 1990) 48 min., $12.98.

500 HOME RUN CLUB. 14 greatest home run hitters, hosted by Bob Costas and Mickey Mantle. 55 min., $19.95.

FUTURE LEGENDS OF BASEBALL. Basic profiles of Boggs, Clemens, others, excerpted from Great Sports Legends series. (RVI) 47 min., $9.95.

GOLDEN GREATS OF BASEBALL. Highlights of baseball's all-time greats. (RVI) 30 min., $14.95.

GOLDEN GREATS OF BASEBALL: *Pitchers.* Highlights of baseball's all-time great pitchers. (RVI) 30 min., $14.95.

GRAND SLAM. Dick Schaap and Billy Crystal host a tribute to baseball. (VID, 1991) 98 min., $19.98.

GREATEST HOME RUN HITTERS. Includes interviews with Mantle, Stargell, Schmidt, et al. (RVI) 30 min., $9.95.

INTRODUCTION TO BASEBALL CARD COLLECTING. Basics of collecting narrated by Bobby Valentine; also sold in a deal with 100 baseball cards from the late 1980s. (JCI) 28 min., $9.95.

1950s BASEBALL DREAM TEAM. Sold as two-tape set with Fantastic Baseball Bloopers. 30 min., $9.95.

Instructionals

ART OF HITTING. Hitting coach Vada Pinson demonstrates everything from power hitting to bunting. 59 min., $19.95.

ART OF HITTING .300. Hitting guru Charlie Lau's classic book on video. 50 min., $19.99.

BASERUNNING BASICS WITH MAURY WILLS. Tips on stealing, sliding, leading off, etc. by the man who brought the steal back to baseball. 60 min., $19.95.

BASEBALL BUNCH: Fielding. Johnny Bench with Ozzie Smith, Graig Nettles and Gary Carter on catcher's role as field general and on fielding grounders. (TW, 1986) 54 min., $14.98.

BASEBALL BUNCH: *Hitting*. Johnny Bench with Lou Piniella, Jim Rice, and Ted Williams on hitting techniques. (TW, 1986) 60 min., $14.98.

BASEBALL BUNCH: *Pitching*. Johnny Bench with Dan Quisenberry, Tom Seaver and Tug McGraw on pitching mechanics. (BA, 1986) 59 min., $14.98.

BASEBALL WITH ROD CAREW. 60 min., $29.95.

BASEBALL MASTERS SERIES, $14.95 each.

CONDITIONING AND BASERUNNING. Jerry Kindall on conditioning and Al Kaline on baserunning. 40 min.

FIELDING. George Kell and Frank Quilici on infield play, Al Kaline on outfield play. 40 min.

HITTING. Al Kaline and George Kell on hitting. 25 min.

PITCHING. Jerry Koosman on pitching. 25 min.

BASEBALL OUR WAY. Instruction from Eric Davis, Wally Joyner, Tommy Lasorda, and others. 90 min., $29.95.

BASEBALL SKILLS AND DRILLS WITH DR. BRAGG STOCKTON. $39.95 each.

  VOL. 1: COACHING PSYCHOLOGY
  VOL. 2: DEFENSIVE SKILLS BY POSITION
  VOL. 3: FUNDAMENTALS OF BASERUNNING
  VOL. 4: FUNDAMENTALS OF FIELDING
  VOL. 5: FUNDAMENTALS OF HITTING
  VOL. 6: FUNDAMENTALS OF PITCHING

BASEBALL THE PETE ROSE WAY. Rose instructional on batting, fielding, baserunning. (EM, 1986) 60 min., $19.98.

BASEBALL THE RIGHT WAY, $9.95 each.

1. FIELDING FOR KIDS. Fielding Fundamentals with Bud Harrelson. 30 min.
2. HITTING FOR KIDS. Hitting Fundamentals with Bill Robinson. 30 min.
3. PITCHING FOR KIDS. Pitching Fundamentals with Mel Stottlemyre. 30 min.

BASEBALL THE YANKEE WAY. Mantle, Maris, Ford, Howard, Richardson, Downing. (1964) 45 min., $19.95.

BASEBALL TIPS. Former Yankees Mickey Mantle, Whitey Ford, and Phil Rizzuto on hitting, pitching, and fielding. (CBS, 1986) 90 min., $19.98.

DO IT BETTER: Baseball Basics Sports Video Series; hosted by California Angels' coaches Marcel Lachemann and Vincent Capelli. $24.95.

DOCTOR'S PRESCRIPTION FOR THE PITCHER: *A Step-by-step Total Body Conditioning Program Medically Designed to Improve Pitching Performance.* Two-video set; hosted by Dr. Arthur Pappas, with Dennis Eckersley. $29.95.

GEORGE BRETT'S SECRETS OF BASEBALL. Guide to hitting. 25 min., $19.95.

JOSE CANSECO'S BASEBALL CAMP. Baseball's first 40/40 man on fundamentals. (IVE, 1989) 60 min., $19.95.

COACHING CLINIC. A coaching video on baseball by Major League Baseball. 62 min., $19.95.

DODGERS WAY TO PLAY BASEBALL. All aspects of the game, narrated by Vin Scully. 90 min., $19.95.

TONY GWYNN'S KING OF SWING. How to hit by one of the best. (PHV, 1991) 35 min., $19.95.

TONY GWYNN'S PLAY TO WIN. Sharpening your fielding,

baserunning, etc. (PHV, 1991) 35 min., $19.95.

HITTING: *Getting the Feel of It.* U. of Iowa Coach Tom Petroff reveals his revolutionary "see & feel method" of hitting. $29.95.

HITTING MACHINE: *Featuring the Star System (Stride, Trigger, Assemble, Release).* 125 advanced drills and techniques by Oklahoma State Coach Gary Ward. 87 min., $39.00.

HITTING WITH HARRY "THE HAT" WALKER. Fundamentals by a former N.L. batting champion. $29.95.

HOW TO PLAY BETTER BASEBALL. Mike Schmidt hosts, with Dale Murphy on hitting, Tim Raines on baserunning, and Frank White on infield play. 47 min., $9.95.

DICK HOWSER'S BASEBALL WORKOUT. Combines Volumes 1 and 2 (see below) by the highly respected late manager. 120 min. $29.95.

> VOL. 1: PITCHING, HITTING AND INFIELD PLAY. 60 min.
> VOL. 2: CATCHING, BUNTING, OUTFIELD PLAY, BASERUNNING AND SLIDING. 60 min.

INFIELD TECHNIQUES AND CATCHING WITH BUD MIDDAUGH. Techniques, drills, etc. from U. of Michigan head coach. 45 min., $39.95.

LITTLE LEAGUE: *How to Hit and Run.* With Coach Bragg Stockton. 30 min., $9.99.

LITTLE LEAGUE: *How to Pitch and Field.* With Coach Bragg Stockton. 30 min., $9.99.

LITTLE LEAGUE'S OFFICIAL HOW-TO-PLAY BASEBALL BY VIDEO. An in-depth instructional on the basics of baseball. (MA, 1989) 70 min., $19.95.

MASTER THE SECRETS OF THE HITTING MACHINE. Detailed instruction by Oklahoma State Coach Gary Ward. 55 min., $39.00.

MICKEY MANTLE'S BASEBALL TIPS FOR KIDS OF ALL AGES. Mick, Whitey Ford & Phil Rizzuto cover batting, pitching, and fielding basics. 40 min., $19.98.

LOU PINIELLA's WINNING WAYS. The Reds manager discusses fundamentals, along with Dave Magadan and Chris Sabo. (CMV, 1991) 51 min., $19.95.

PITCHING ABSOLUTES. Tom House on pitching grips, conditioning, and theory. $39.95.

PITCHER'S FIELDING PRACTICE. Texas Rangers pitching coach Tom House provides instruction on holding runners, turning two, fielding bunts, covering bases, comebackers, and coaching techniques. 55 min., $39.95.

PITCHING MECHANICS: *Problem Recognition and Solutions.* Sequel to "Pitching Absolutes;" Tom House on biomechanics, trouble shooting, throwing mechanics, drills, etc. 93 min., $39.95.

PITCHING STRATEGIES AND TACTICS. A day in the life of a pitcher; conditioning, training, and routines used by the pros. $39.95.

PITCHING WITH BUD MIDDAUGH. Grips, drills, techniques, etc. from U. of Michigan head coach. 45 min., $39.95.

PLAY BALL SERIES. $9.95 each.

1. CATCHING WITH LANCE PARRISH. 30 min.
2. HITTING WITH AL KALINE. 30 min.
3. PITCHING WITH ROGER CRAIG. 30 min.

PLAY BALL WITH REGGIE JACKSON. Tips and drills on batting, fielding, baserunning, and pitching. 30 min., $9.95.

PLAY BALL WITH MICKEY MANTLE. Mick, Gary Carter & Tom Seaver present inside tips for kids. (CBS, 1987) 90 min., $19.98.

PROFESSIONAL HITTER. Michigan State U. Coach Rob Ellis presents an easy-to-follow outline format on hitting, including slow motion, stop action, hitting psychology, etc. 100 min. $49.95.

PROFESSIONAL SPORTS TRAINING FOR KIDS. Hitting with Ken Griffey, Jr. Hitting clinic with the Mariners' superstar, plus conditioning workouts for hitting with Olympic trainer Pete Schmock. 47 min., $19.95.

PETE ROSE ON WINNING BASEBALL. Claude Osteen on pitching and Sonny Ruberto on catching. 55 min.

SCIENCE OF PITCHING. Pitching Coach Wes Stock demonstrates the basics. 60 min. $19.95.

SIMPLIFIED FUNDAMENTALS OF HITTING AND BUNTING. Southern Illinois coach Richard "Itch" Jones. 45 min., $39.95.

SPORTS CLINIC: BASEBALL. Instructions on pitching, batting, fielding, and baserunning hosted by Dick Williams. (SCU, 1987) 80 min., $19.99.

SPORTS TEACHING VIDEO: BASEBALL. All-around basics with former major leaguer Don Kessinger. 25 min., $12.95.

SPORTS TRAINING CAMP: BASEBALL. Host Graig Nettles guides the viewer through "how-to" game action. 60 min., $19.95.

STEVE GARVEY'S HITTING SYSTEM. Instruction by the former Dodgers and Padres star. (TW) $19.95.

TEACHING KIDS BASEBALL WITH JERRY KINDALL. U. of Arizona coach on teaching beginning baseball to youngsters. (ESP, 1988) 75 min., $29.95.

TEACHING THE MECHANICS OF THE MAJOR LEAGUE SWING. Based on actual swing mechanics of top major league hitters. $39.95.

UMPIRING BASEBALL: *The Third Team on the Field.* Individual responsibilities of home plate and base umpires. 17 min. $29.95.

Bloopers, Humor

BASEBALL FUN AND GAMES. Baseball bloopers hosted by Joe Garagiola. (VID, 1987) 60 min., $14.98.

BASEBALL FUNNIES: *A Hilarious Look at Baseball.* Compilation of plays, fantastic, famous or funny. (SIM, 1988) 30 min., $9.95.

BASEBALL FUNNY SIDE UP. Humorous Moments in baseball hosted by Tug McGraw. (1987) 45 min., $9.95.

BASEBALL LAUGHS, GAFFES & GOOFS. With Jay Johnstone, Tommy Lasorda, Dallas Green, and others. (RVI) 30 min., $9.95.

BASEBALL'S OFFICIAL BALLPARK BLOOPERS. Narrated by Warner Fusselle. $14.95.

FANTASTIC BASEBALL BLOOPERS. Sold as a two-tape set with 1950s Baseball Dream Team. 30 min., $9.95.

PRO BASEBALL'S FUNNIEST PRANKS. The greatest practical jokes of the diamond. 30 min., $9.95.

SUPER DUPER BASEBALL BLOOPERS. Collection of the bizarre. (MLB/BLK, 1989) 49 min., $19.95.

See the notes in the Introduction relating to Blockbuster Video's contract with Major League Baseball.

BASKETBALL

I must confess the delights of basketball have never been clear to me. This no doubt reflects some basic flaw in my character. *Everybody* loves hoops but me. While all around me, my friends and neighbors are watching "March Madness" on the tube, I'm likely to opt for a *Green Acres* rerun. At most, I may tune into a championship game in the final minutes to see who won, but even then I usually end up only hearing the result.

Perhaps it all goes back to my grade school days. The only basketball courts in my neighborhood were monopolized by the big kids, who snickered and sneered when a chubby third-grader threw up an air ball. Had some kindhearted sixth-grader taken me by the hand and said, "Here, Fatso. This is how it's done," my attitude toward basketball might have turned out differently.

Bottom line: the thought of viewing all those basketball videos put me in a tizzy. I couldn't delude myself into believing I could be in any way fair in assessing their worth.

Fortunately, my friend John McNamara, while otherwise admirable in many ways, *likes* basketball. A lot.

I prevailed upon him to save me from a fate almost worse than death by taking over this section for me. The following is his 360-view of basketball videos.

During the 1980s, there were plenty of changes in the world around us. There was the Reagan revolution, another Russian revolution, and the advent of *Entertainment Tonight*, just to name a few.

However, in the home entertainment industry the biggest revolution in the past ten years involves the growing sophistication and popularity of home video cassette recorders. One estimate is that today in the United States, more than 60 percent of the homes now count VCRs among their home entertainment systems.

And, if asked to name the biggest change in spectator sports during the last decade, most would cite the mushrooming popularity of college and professional basketball. Even if you don't know how to program your VCR, you probably know a good deal about Michael and Magic and Larry.

Considering where the National Basketball Association is today in terms of popularity, it's hard to imagine that so much has changed in a little more than 10 years. Back in 1979–80, when rookie Magic Johnson led the Lakers to the first of their five championships in a decade, the finals weren't even on during prime time. That's right. CBS thought so little of the draw of these games (featuring Johnson, Kareem Abdul-Jabbar, and Julius Erving, among others), or thought so highly of its regular weeknight programming, that the finals were actually shown on tape delay on network television.

Less than 10 years later, CBS would finally cash in on the sport's burgeoning popularity and would televise a regular-season game in prime time, something that hadn't been done in the previous 20 years.

Yes, much has changed. Pro and college basketball is all over the airwaves and the video stores. The main reason for this is probably the good folks at NBA Entertainment, Inc. The league has done a marvelous job of marketing and its efforts in home video is no exception.

For years, NBA Entertainment has linked up with CBS/Fox Video to market all kinds of videos on that subject. Making this much easier was the fact that CBS had exclusive network television rights to the NBA, its playoffs, as well as the NCAA basketball tournament.

Now that NBC has taken over the network broadcast of league games, it's too early to tell what kinds of basketball home videos might be coming. But as long as Michael Jordan and Larry Bird are around, there will be enough highlights to go around.

Team Highlights

Besides the expected works—NBA championship and Final Four, Michael Jordan and Magic Johnson—there are videos for the more casual fan, featuring things like fantastic finishes on one tape. Two "Dunks and Bloopers" videos are available for folks who like their NBA action on the lighter side. If it happened during the 1980s, somewhere someone's got it on tape and is willing to sell it.

How else could you explain the existence of a Charlotte Hornets video, recapturing on film all the highlights from that franchise's inaugural season? The boys in blue had great uniforms (courtesy of Alexander Julian) and sold more clothing and souvenirs that any team in the NBA that year, but the franchise won a measly 20 games. Maybe that's why their video, *The Season That Made Charlotte Shout*, is only about 20 minutes long. I bet not even Kelly Tripucka has a copy.

Fortunately for serious basketball fans, the other team highlight videos are a little more substantive. For any purple-and-gold-blooded Laker fan, the tape documenting the team's consecutive championships is a must. Perhaps no team in basketball history ever ran as exciting and spectacular a fastbreak as the Lakers have. If you don't believe that, catch an eyeful of this video, *Back to Back*. It's got Magic driving and dishing, Worthy swooping to the hoop, plus Kareem's majestic sky hooks. Now that Magic Johnson has retired, the Lakers won't ever be the same. If you want a souvenir of "The Team of the '80s" that will stay with you, this is the one you should buy.

Similarly, Pistons fans can have their own memento of their team's rise to prominence. There are tapes from the season the Pistons lost in the finals, and another for the next year, when they knocked off the Lakers. Both feature some great candid footage of the team on the road, joking around in the locker room, as well as their hardnosed play on the court. Oddly, these two collections reinforce the team's "bad boy" image on the court, while contradicting it with scenes of the guys off the playing floor. It appears that if you got to know them, Laimbeer and Rodman wouldn't seem like such bad guys, after all. Maybe.

Still, some of the selections for team highlight videos do seem a little curious. For example, why is there a video available about the Dallas Mavericks' 1987–88 season? The Mavericks, who enjoyed tremendous success after entering the league in 1980, reached the conference finals that season, getting eliminated by the Lakers. I love watching Rolando Blackman play, but could anybody name the other four starters on this team? Does anyone really care who finished second in the Western Conference playoffs in 1988? If you have the answers to these questions, I'll give you my copy of the Mavericks video, okay?

Similarly, there's a video entitled *Higher Ground: Chicago Bulls 1987/88 Season*. Anyone want to hazard a guess as to why this video was created? Anyone want to guess who's on the cover of the video? I'll give you a hint. It's not Doug Collins.

The tape chronicled the exploits of the Chicago Bulls that season—a pretty good team that won 50 games and lost in the second round of the playoffs. Big deal. Back in the early '70s, the Bulls had plenty of teams that could do that, but you never saw Bob Love or Chet Walker captured forever on videotape, didya?

This was the rookie season for both Scottie Pippen and Horace Grant, and with the nonpareil Jordan leading the way, it was clear that these Bulls were headed somewhere, someday. But we hear raves for players like Charles Oakley and Rory Sparrow, who had nothing to do with the team that captured the championships three seasons later. Doug Collins was coaching the Bulls that year, and even he was gone by the time the franchise won its first NBA title.

This is just one example of Jordan overkill. Of course he's the most spectacular player in the history of the sport, a man who has brought the game to new heights. But every time I turn around, I bump into this guy (and he never gets called for the foul).

This video was supposed to be about the Bulls' season, and we get a full 20 minutes of highlights from the All Star Game, which was played in Chicago that year. Of course, it just so happened the Jordan won the slam-dunk contest, and we see it all spectacularly captured in super slow motion.

But we've seen it all before. There's *Michael Jordan: Come Fly With Me,* and *Michael Jordan's Playground* for the Jordan Connoisseurs. A lot of the same dunks and highlights show up on all three videos.

And, keep in mind, all three of these tapes were released long before Jordan and the Bulls ever won an NBA title. Who knows how many taped highlight packages are still to come? By that time, Jordan will become like Phil Collins or Michael Caine. The thrill will be gone, through overexposure.

Not even Jordan devotees need to have all three tapes. If you're gonna buy one, go for *Playground* or *Come Fly With Me*. The former highlights Jordan's overall game a bit more, showing some of his defensive abilities as well as the usual jumpshots and breakaway dunks. *Come Fly With Me* offers the kind of offensive pyrotechnics the title suggests.

One team highlight video that every basketball fan should at least rent is the one detailing the history of basketball's clown princes, the Harlem Globetrotters. There are plenty of memorable highlights in this tape, plus some interesting history. Did you know that during World War II, when able-bodied males were at a premium, that the team actually had a white player? Or that the team originated in Chicago, and founder Abe Saperstein merely used "Harlem" in the name to let everyone know that the team was all-black?

General Highlight Videos

Which brings us to another kind of highlight video. For lack of a better term, we'll call these general highlight videos. These tapes examine a particular aspect of the game, such as bloopers, "awesome endings," slam dunks, or whatever.

Happily, the news in this type of basketball video is almost universally good. For basketball video buyers who have grown up with the NBA and didn't jump on the bandwagon until later, there's a lot to enjoy here.

If you want to start at the beginning of the alphabet, then try *Awesome Endings*. This collection brings back to life some of the great last-second finishes in league history—including a couple of games I had forgotten about. Just about every great player is in here—Magic Johnson, Larry Bird, Jerry West, Bernard King, John Havlicek. There is a good selection of games from the '60s, '70s, and '80s. The narration and soundtrack emphasize the drama of these games, and it's certainly the equal of anything NFL Films had ever put together.

Also on this level is *Showmen: The Spectacular Guards of the NBA*. The Magic and Michael stuff can also be seen elsewhere, but the inclusion of some other footage is what makes this tape so interesting. If you've been following the game for a few years, then the moves and magic of players like Pete Maravich, Tiny Archibald, and Earl Monroe will be that much better seen again. If you're really interested in learning more about the league's history or if you just want to see some of the great games and players once again, then check out the *History of the NBA*, which is narrated by Knick coach Pat Riley.

Two blooper videos have been released through CBS/Fox, which is really about one and a half too many. The footage in these videos is great, but we get an overdose of broadcaster Marv Albert and funnyman/ex-coach Frank Layden. I've heard Layden as an after-dinner speaker, and the man really is funny. But not on these tapes. He tries to make himself into basketball's Bob Uecker or Rodney Dangerfield, and he succeeds only in getting in the way. You only need one of these tapes—either the original *Dunks and Bloopers* or the sequel.

Historical Highlights

If college basketball is more your passion, then *Great Moments in College Basketball* and *Final Four: The Movie* are a couple of titles you could check out. All the great coaches, places, and players are here.

This brings us to one problem with the recent basketball video explosion. Because the game became so popular in the last decade, a lot of the tapes are geared toward active players. Although there have been great teams and great moments long before Larry and Magic came along, many of them aren't on film. For example, there is no definitive Boston Celtics team history video. The team has won 16 league titles, enjoyed success unparalleled in the sport, yet there's no film monument to the franchise.

There is a commemorative video for the title winning teams in 1981, 1984 and 1986 that featured Bird, McHale, and Parish, but none for any of the other title-winning Celtic teams, thus bypassing such figures as Russell, Cousy, Havlicek, and Auerbach.

I think it would be great to have a collection of the Celtics-Lakers finals, dating back to the West-Baylor Laker teams of the 1960s. Over the years, those two teams have staged some epic battles for the championship, but so far, no one has thought to compile them on film.

On the other hand, there is a video, *The New York Game*, which offers 40 years of team history on the New York Knicks, including the championship teams of 1970 and 1973.

Fortunately for the NBA archivists among us, there are some great tapes highlighting individual players. The Greatest Sports

Legends series features biographical videos on such figures as Red Auerbach, Wilt Chamberlain, Julius Erving, and Kareem Abdul-Jabbar.

All of these videos are entertaining, but an aspiring player would be hard pressed to learn much about how to play the game from watching them. The best basketball players in the NBA are so good that it's tough to figure out just how they do what they do.

Instructional Videos

For that reason, instructional videos were developed. The VCR might be the perfect piece of equipment for young players, because it enables them to watch the videos and run back the parts they need to see again, making it much easier to pick up the nuances of shooting, dribbling, and defense.

And there are plenty of instructionals to choose from. Although there's no Celtics highlight video, there is an instructional featuring Bird, Auerbach, and McHale. As expected, the Celtic video is excellent for its emphasis on the fundamentals of the game. Magic Johnson's *Put Some Magic In Your Game* is also very good, and the game's premier showman makes it fun to watch and learn, thanks to his infectious manner and that famous smile. Spud Webb also has his own instructional/highlight video out, as does Julius Erving. Erving's video includes some great highlights and some of his tricks of the trade that will help the young player especially.

There is a series of instructional videos created by the repeat champion Laker teams from the late 1980s, but those aren't as good. The tapes try to cover too much ground too quickly—a lot like the fabled Laker fast break of those years.

If you can find it, try to get the Sports Clinic basketball video, which features Bill Walton, former UCLA teammate Greg Lee, and former Bruin player and coach Walt Hazzard. This one lasts for 80 minutes, and delivers its message with plenty of repetition and review. It covers everything from defensive stances to the pick and roll. There's not a more thorough video on the market.

For the advanced player or even a coach looking for a tape that outlines some more sophisticated drills and skills, Rick Pitino's video series is very good. Pitino, who coached Providence College, the Knicks, and Kentucky, created a four tape series—two on of fense, two on defense. These are designed for the more advanced player.

And, if coaches/fans want to see how the professional coaches do it, then *Coach to Coach* would be a good one to get. That tape features Pat Riley, Chuck Daly, and Billy Cunningham, among others, explaining how they get the best basketball players in the world to play to their abilities.

Basketball Videos

Prices on highlight videos vary a great deal, but most will run between $9.95 for closeouts to $24.95 for the newer, full-priced videos. Within that range, the great majority of the videos will be priced between $14.95 and $19.95. Many instructionals are in the $30 to $50 range. As always, it pays to shop around for the best price.

College Highlights and History

ACC BASKETBALL: *35 Years of Excellence.* Chronicle of the top college basketball conference in the country, showcasing all the great Carolina players, from Cunningham to Jordan, plus Ralph Sampson, David Thompson, and many others. Narrated by Jim Lampley. (RAY) 60 min., $14.95.

FINAL FOUR: THE MOVIE. This one has it all, including footage of almost every final since the NCAA tournament began in 1939; plus interviews with key players and coaches. (JCI) 90 min., $19.95.

GREAT MOMENTS IN COLLEGE BASKETBALL. A perfect companion for the NCAA tournament video; covers all the highlights, from the point shaving scandals of the early 1950s to the UCLA dynasty. (VES) 48 min., $19.98.

(Kansas) MARCH THROUGH MADNESS: *1990–91 Kansas Basketball.* Highlights of the Jayhawks' run to the national championship game. 60 min., $19.99.

(Kentucky) UNKNOWN, UNRANKED... UNFORGETTABLE: *Kentucky Basketball 1990–91*. 58 min., $19.99.

(North Carolina) TAR HEELS ON TAPE: *The Team That Wouldn't Quit*. Season highlights of the surprising 1989–90 North Carolina team. 60 min., $19.95.

(North Carolina) TAR HEELS ON TAPE. Official UNC basketball recruiting and highlight film, covering 1990–91 season plus NCAA. 45 min., $24.95.

OFFICIAL NCAA CHAMPIONSHIP VIDEO: *1990*. The Runnin' Rebels win the championship. (CBS) 60 min., $19.98.

OFFICIAL NCAA CHAMPIONSHIP VIDEO: *1991*. Duke's revenge. The Blue Devils finally win their title, beating defending champ UNLV in the semifinals. (CBS) 60 min., $19.98.

TIME OUT, BABY! DICK VITALE'S COLLEGE HOOP SUPERSTARS. Vitale picks the all-time greats in college basketball from the 1950s through the '80s, plus his all-time NCAA team. You won't have to turn up the volume on this one. (ESP) 30 min., $9.95.

Pro Basketball Highlights: General

BASKETBALL DREAM TEAM. Featuring Oscar Robertson, Jerry West, Elgin Baylor, Julius Erving, and Kareem Abdul-Jabbar. $14.99.

BASKETBALL'S AMAZING RAMS, SLAMS AND JAMS. Sold as "double feature" with FANTASTIC BASKETBALL BLOOPERS. 30 min. each, $9.95.

THE DYNASTY RENEWED. A look at the playoffs and NBA Finals from the 1980–81 season. Boston beats the Houston Rockets in the first of three Celtic titles in the 1980s, also the first with the Bird-Parish-McHale frontline; Brent Musburger narrates. 30 min., $19.98.

GOLDEN GREATS OF BASKETBALL. Highlights 11 of the best players ever in the game; newer fans can get a look at players like Bob Pettit and Elgin Baylor from the earlier days of the league. 45 min., $14.95.

HARLEM GLOBETROTTERS: *Six Decades of Magic.* Covers everything from the team's humble beginnings as a barstorming squad out of Chicago to the worldwide phenomenon they are today. Louis Gossett, Jr. narrates. (FRI) 60 min., $19.95.

HISTORY OF THE NBA. A treasure trove of highlights and interviews, featuring Russell, Cousy, Mikan, Wilt, and the rest. A must for the basketball archivist. Knick coach Pat Riley narrates. (CBS) 60 min., $19.98.

LEGENDARY BIG MEN. Another of the Greatest Sports Legends series, spotlighting Jabbar, Chamberlain, Russell, Mikan, and Willis Reed. (RVI) 30 min., $9.99.

MADISON SQUARE GARDEN'S ALL-TIME GREATEST BASKETBALL FEATURING THE HARLEM GLOBETROTTERS. Early film of the Globetrotters. 30 min., $14.95.

NBA ALL-STAR WEEKEND 1986. Slam dunk and three-point shooting contests, game highlights. $19.98.

NBA AWESOME ENDINGS. Game-winning shots and last minute heroics. (CBS) 48 min., $14.95.

NBA SHOWMEN: *The Spectacular Guards.* More rare footage of some of the great guards of the past. If you like Magic's moves and Drexler's shots, then take a look at the clips of Pete Maravich, Nate Archibald, and Earl Monroe. (CBS) 60 min., $14.98.

NBA SUPERSTARS. Features highlight footage of eight top players set to pop music tunes. (CBS) 48 min., $14.98.

SUPER SLAMS OF THE NBA. Delivers more than 400 dunks in 40 minutes, including the best of Darryl Dawkins, Julius Erving, and Spud Webb, among others. (CBS) 40 min., $14.95.

Pro Basketball Highlights: Teams

ATLANTA HAWKS 1986–87: *Basketball's Air Force.* (CBS) $19.98.

BOSTON CELTICS 1986–87: *Home of the Brave and Sweet 16.* A double tape with highlights from the Celtics' 1987 team that lost in the finals to the Lakers, plus a look at the 1986 championship team. (CBS) 60 min., $19.98.

CHARLOTTE HORNETS 1988–89: *Hornet Hysteria.* The franchise led the league in attendance while winning only 20 games, but the highlights were few on the court. (CBS) 25 min., $24.98.

CHICAGO BULLS 1987–88: *Higher Ground.* The first season for Horace Grant and Scottie Pippen. And you-know-who flies through. (CBS) 45 min., $19.98.

CHICAGO BULLS 1990–91: *Learning to Fly.* The Bulls' first ever championship. Although Jordan was obviously the star of this team, the key player in the finals was slow-footed John Paxson. (CBS) 45 min., $19.98.

DALLAS MAVERICKS 1987–88: *Bouncing Back.* Game highlights, plus a humorous golf outing and an intense speech by coach John McLeod. (CBS) 60 min., $19.98.

DETROIT PISTONS 1987–88: *Bad Boys.* Chronicles the Pistons' first trip to the NBA Finals, where they lost to the Lakers. Includes lots of behind-the-scenes stuff. (CBS) 60 min., $19.95.

DETROIT PISTONS 1988–89: *Motor City Madness.* More Pistons stuff, this time after the team finally won the NBA championship, knocking off the Lakers. (CBS) 50 min., $9.98.

DETROIT PISTONS 1989–90: *Pure Pistons.* Detroit becomes a team for the ages, winning a second straight title. Includes historical footage of the early days of the franchise, including the beginnings in Ft. Wayne, Indiana, and past stars like Bob Lanier and Dave Bing. (CBS) 48 min., $9.98.

HOUSTON ROCKETS 1986–87: *Hanging Tough.* The Twin Tower offense of Coach Bill Fitch in game action. (CBS) 45 min., $19.95.

LOS ANGELES LAKERS 1986–87: *World Championship Drive for Five.* (CBS) 45 min., $19.98.

LOS ANGELES LAKERS 1987–88: *Back to Back.* The great Laker teams of the late 1980s; features Magic dishing off, Worthy swooping, and Pat Riley stylin'. Long-time Laker broadcaster Chick Hearn narrates. (CBS) 48 min., $9.98.

MIAMI HEAT 1988–89: *The Dream Catches Fire.* A look at the inaugural season of the Miami Heat. (CBS) 30 min., $24.98.

MINNEAPOLIS LAKERS 1950–51: *Basketball Fundamentals* and *1952–53: Meet the Champs.* Game footage plus demonstration of plays. 48 min., $29.98.

NEW YORK'S GAME. History of the New York Knicks, highlighting the championship teams from 1970 and 1973, plus lots of Patrick Ewing highlights for the younger crowd. (CBS) 45 min., $19.98.

NBA Playoff Highlights

1980: THAT MAGIC SEASON. Lakers vs. Philadelphia. (CBS) $9.98.

1981: THE DYNASTY RENEWED. Boston vs. Houston. (CBS) $9.98.

1982: SOMETHING TO PROVE. Lakers vs. Philadelphia. (CBS) $9.98.

1983: THAT CHAMPIONSHIP FEELING. Philadelphia vs. Lakers. (CBS) $9.98.

1984: PRIDE AND PASSION. Boston vs. Lakers. (CBS) 45 min., $9.98.

1985: RETURN TO GLORY. Lakers vs. Boston. (CBS) $9.98.

Highlights: Individuals

KAREEM ABDUL-JABBAR. Greatest Sports Legends. (RVI) 30 min., $6.98.

KAREEM: *Reflections from Inside.* Put together near the end of Kareem's career, and he seems much more forthcoming and reflective than he is in the GSL video. He narrates this video himself, and gives us a look at the neighborhood where he grew up, including the playground courts where he first learned the game. (CBS) 60 min., $24.98.

RED AUERBACH. Greatest Sports Legends. (RVI) 30 min., $6.98.

LARRY BIRD: *Basketball Legend.* (1992) 60 min., $9.95.

WILT CHAMBERLAIN. Greatest Sports Legends. (RVI) 30 min., $6.98. (Also available as Doubleheader: WILT CHAMBERLAIN / KAREEM ABDUL-JABBAR. Greatest Sports Legends. (RVI) 60 min., $14.95.

JULIUS ERVING. Greatest Sports Legends. (RVI) 30 min., $6.98.

JOHN HAVLICEK. Greatest Sports Legends. (RVI) 30 min., $9.95.

MAGIC JOHNSON: *Always Showtime.* Career highlights; one of those *Sports Illustrated* bonuses that will eventually find its way into the stores. (1992) 60 min., price n/a.

MICHAEL JORDAN: *Come Fly with Me* The most spectacular of the Jordan videos, at least if you like to watch him dunk. There's plenty of tongue-wagging in this tape. (CBS) 45 min., $19.98.

OSCAR ROBERTSON. Greatest Sports Legends. (RVI) 30 min., $9.95.

BILL RUSSELL. Greatest Sports Legends. (RVI) 30 min., $9.95.

JERRY WEST. Greatest Sports Legends. (RVI) 30 min., $9.95.

Instructionals

> Be there a hoops coach so brain dead
> Who never to himself has said,
> "My bankbook will look oh-so pretty, oh,
> *After I make my instructional video!"*

ARKANSAS RAZORBACK ATTACK. Nolan Richardson on Arkansas' "Attack" system on both offense and defense. $49.95.

BASKETBALL COACHES CORNER TEACHING TAPE. Four top women's coaches —Andy Landers of Georgia Tech, Theresa Grentz of Rutgers, Jody Conrad of Texas and Pat Summitt of Tennessee— teach fundamentals of women's basketball. 90 min., $39.95.

BASKETBALL IN THE FAST LANE. Paul Westhead details his explosive fast break system of basketball. $49.95.

BASKETBALL FOR THE '90S: *Beginning Basketball with Bibby.* For the beginning player, with Coach Henry Bibby. 57 min. $39.95.

BASKETBALL FOR THE '90S: *Building a Championship Defense.* Stanford women's coach Tara VanderVer reviews her championship defensive system. 55 min., $39.95.

BASKETBALL FOR THE '90S: *Building a Championship Offense.* Stanford women's coach Tara VanderVer on her championship offensive system. 55 min., $39.95.

BASKETBALL FOR THE '90S: *Developing the Big Man.* Lute Olson teaches procedure, psychology, physical, and skill developers. 57 min., $39.95.

BASKETBALL FOR THE '90S: *Developing the Perimeter Player.* Lute Olson teaches philosophy, ball-handling, and shooting skills. 50 min., $39.95.

BASKETBALL FOR THE '90S: *The "Shark" on Defense.* UNLV's Jerry Tarkanian teaches defensive techniques and drills, including the "amoeba" defense. 52 min., $39.95.

BASKETBALL FOR THE '90S: *The "Shark" on Offense.* UNLV's Jerry Tarkanian teaches individual and team offensive techniques and drills. 55 min., $39.95.

BASKETBALL WITH BILL FOSTER AND GAIL GOODRICH. Dribbling, jumping, and shooting with the pros. $19.95.

BASKETBALL WITH HUBIE BROWN. Coach Brown's offensive and defensive techniques. 90 min., $39.95.

BECOMING A BASKETBALL PLAYER SERIES. Springfield College Coach Hal Wissel teaches basketball based on his popular book. 20 min., $19.95 each.

    BALL HANDLING. Passing, catching and dribbling.

    DEFENSE AND REBOUNDING. Developing quickness, jumping, rebounding, tipping.

    OFFENSIVE MOVES. Shooting, passing, driving.

    OFFENSIVE MOVES OFF THE DRIBBLE. Speed dribble, change of pace, reverse.

    SHOOTING. Layups to three-pointers.

BIG MAN CAMP. Bill Walton on offense, defense, and rebounding for centers and forwards. 33 min., $29.95.

BOB HURLEY'S CHAMPIONSHIP BASKETBALL SERIES. By the coach of St. Peter's of Jersey City. $24.95 each.

    VOL. 1: OFFENSE.

    VOL. 2: DEFENSE.

BOBBY KNIGHT'S BASKETBALL CLINIC SERIES. Conducted by Indiana's legendary coach. 120 min., $49.95 each.

    VOL. I: OFFENSIVE AND DEFENSIVE DRILLS.

    VOL. II: MOTION OFFENSE.

BOEHEIM ON BASKETBALL. Syracuse coach teaches techniques for younger players. 45 min., $29.95.

BOMBS AWAY: *Let's for 3.* Ron Righter on the 3-point shot. 45 min., $49.95.

COACH TO COACH. NBA coaches talk about what they do, how they plan, and the unique requirements for coaching the best players in the world, featuring Billy Cunningham, John McLeod, Chuck Daly and Pat Riley. (CBS) 60 min., $39.95.

DO IT BETTER: BASKETBALL. ASICS Sports Video Collection; with coaching legend Pete Newell. 30 min., $24.95.

DO IT BETTER: WOMEN'S BASKETBALL. ASICS Sports Video Collection; with Long Beach St. coach Joan Bonvicini. 30 min., $24.95.

Dr. J's BASKETBALL STUFF. Basketball tips plus great highlights. (CBS) 60 min., $19.95.

DON MEYER BASKETBALL SERIES. Coach of NAIA champion David Lipscomb University. 60 min. except where indicated, $35.00 each.

> **BECOMING A GREAT SHOOTER.**
>
> **CHAMPIONSHIP APPROACH TO STRENGTH TRAINING AND CONDITIONING.** 50 min.
>
> **CHAMPIONSHIP PREPARATION FOR GAMES.**
>
> **DEVELOPING YOUR PERIMETER PLAYERS.**
>
> **DEVELOPING YOUR POST PLAYERS.**
>
> **DRILLS FOR TEACHING INDIVIDUAL FUNDAMENTALS AND TEAM DEFENSE.** 85 min.
>
> **DRILLS FOR TEACHING INDIVIDUAL FUNDAMENTALS AND TEAM OFFENSE.** 80 min.
>
> **MATCH-UP ZONE.**
>
> **MOTION OFFENSES:** *Simplified Principles.*
>
> **PRACTICE PLANNING, ORGANIZATION AND ON-THE-FLOOR DEMONSTRATION.** 65 min.
>
> **PRESSURE MAN-TO-MAN DEFENSE:** *A System..*
>
> **TEAM ATTITUDE.**
>
> **TRANSITION GAME.**
>
> **UTILIZING AND DEFENDING THE THREE-POINT SHOT.**

WINNING SPECIAL SITUATIONS. 90 min.

ZONE ATTACK.

FAST BREAK: *The Fundamentals of Championship Basketball.* Starring five L.A. Lakers. (CY) $19.95.

VOL. I: THE BASICS. 70 min., $19.95.

VOL. II: TEAM PLAY. 60 min., $19.95.

HIGH PERCENTAGE BASKETBALL: *Getting the Ball to the Open Man.* Former Pacers' head coach Dick Versace on his system. $49.95.

KNIGHT OF BASKETBALL. Coach Bobby Knight teaches you how to watch basketball. 90 Min., $19.95.

LUTE OLSON'S BASKETBALL SERIES. Arizon's coach teaches team techniques. 36–54 min. each., $49.95 each.

MAN DEFENSE.

MATCH-UP DEFENSE.

PASSING GAME OFFENSE.

MAGIC JOHNSON: *Put Magic in Your Game.* Magic teaches the tricks of his game with highlights of some of his greatest moments. (CBS) 48 min., $19.98.

MAXIMIZING YOUR DEFENSE: *Combatting Today's Offensive Trends.* Ron Righter demonstrates solving problems created by today's offenses. $39.95.

MICHAEL JORDAN'S PLAYGROUND. Michael Jordan helps an aspiring young player make the high school team, thanks to some pointers from the game's biggest star. (CBS) 60 min., $19.98.

MIKE FRATELLO'S 3-POINT PLAY. Strategy and technique of the three-point shot. 40 min., $29.95.

MORGAN WOOTTAN: *Teaching Basketball, Vol. I, Offensive Tips.* DeMatha High School's famed coach. 60 min., $29.95.

NORTH CAROLINA SYSTEM. A look at Dean Smith's system, with Kansas coach Roy Williams on offense and Vanderbilt coach Eddie

Fogler on defense. (WES) 30 min., $14.95.

OFFENSIVE AND DEFENSIVE BASKETBALL TECHNIQUES. Hosted by Coach Jack Ramsey. 30 min., $29.95.

OFFENSIVE LOW POST PLAY. Teaching system and drills from UC at San Diego head coach Tom Marshall. $49.98.

PISTOL PETE'S HOMEWORK BASKETBALL SERIES. Instruction by one of the game's all-time greats. (LAP) 40 min., $29.95 each.

   BALL-HANDLING.

   DRIBBLING.

   PASSING.

   SHOOTING.

POINT GUARD PLAY. Hosted by Cavaliers' head coach Lenny Wilkins. $39.95.

PRESSURE DEFENSE: *A System*. Wisconsin-Green Bay Coach Dick Bennett teaches pressure man-to-man defense. 82 min., $49.95.

REACH FOR THE SKIES WITH SPUD WEBB. The 5'7" star demonstrates techniques; plus game footage (with Dominique Wilkins and Ron Harper). (HNP) 60 min., $14.95.

RED ON ROUNDBALL. Red Auerbach talks with Bob Cousy on backcourt, Pete Maravich on playmaking, Julius Erving on dunking, etc. (BV) 90 min., $19.95.

REEBOK SPORTS LIBRARY: BASKETBALL SKILLS. 30 min., $24.95 each.

   VOL. 1. Foul Shooting by Digger Phelps; Dribbling by Jack Ramsey; Post Play by Leonard Hamilton; Defense by Paul Westphal; Favorite Moves by Roy Hinson and Doc Rivers.

   VOL. 2. Shooting by Jeff Mullins; Shooting Drills by Mike Fratello; Three-Pointers by Stan Albeck; Defense by Johnny Orr; Passing by Gary Williams.

RICK BARRY: *Shooting and Offensive Moves*. ASICS Sports Video Collection. $24.95.

RICK PITINO BASKETBALL SERIES. The well traveled coach has a reputation for being one of the game's smartest. (DR) 60 min., $24.95 each.

> VOL. 1: OFFENSE. Ball handling, one-on-one, shooting.
>
> VOL. 2: OFFENSE. 3-point shots, screens, fast break.
>
> VOL. 3: DEFENSE. Man-to-man, defensive skills and screens.
>
> VOL. 4: DEFENSE. Presure defenses, match-up press, zone traps.

ROLLIE MASSIMINO BASKETBALL INSTRUCTIONAL TAPE. $29.95.

SHOOTING TO WIN: *Fundamentals of Shooting by Steve Alford.* $29.95.

SPORTS CLINIC BASKETBALL. With Walt Hazzard, Bill Walton, and Greg Lee. (SCU) 80 min., $19.95.

SPORTS TEACHING VIDEO: BASKETBALL. Basics of offensive moves, including shooting, dribbling, passing by Paul Westphal. 25 min., $12.95.

SPORTS TRAINING CAMP: BASKETBALL. Walt Hazzard guides you through "how to" game action. 60 min., $19.95.

STAR SHOT. New and revolutionary shooting method featuring B.J. Armstrong and U. of Iowa coaching staff. $39.95.

STEVE ALFORD'S 50-MINUTE ALL-AMERICAN WORKOUT. Build endurance, sharpen shooting skills, etc. 50 min., $29.95.

TEACHING KIDS BASKETBALL WITH JOHN WOODEN. UCLA's legendary coach on teaching beginning basketball. (ESP) 75 min., $29.95.

TEMPLE OF ZONES. Don Casey on zone defense. 38 min., $49.95.

UNLV BASKETBALL SERIES. Hosted by Jerry Tarkanian. $49.95 each.

> UNLV AMOEBA ZONE DEFENSE.

UNLV PRESSURE DEFENSE.

UNLV RUNNING GAME.

WAR ON THE BOARDS. George Raveling on all aspects of rebounding. $49.95.

WINNING BASKETBALL WITH RED AUERBACH AND LARRY BIRD. Larry Bird and a couple of his teammates show you how to play basketball the Celtic way. Excellent tape for beginners. (KOD) 61 min., $19.95.

WINNING WITH THE FLEX OFFENSE. With U. of Iowa assistant coach Ron Righter. 40 min., $49.95.

YOUR BEST SHOT. Bill Russell, Michael Cooper, and others demonstrate shooting techniques. (SBL) 30 min., $29.95.

ZONE OFFENSE. With U. of Iowa head coach Tom Davis. $49.95.

Humor and Bloopers

You'll note that there are fewer blooper videos for basketball than there are for football, or even baseball... but not for lack of material. How about a Sacramento Kings or Dallas Mavericks highlight collection?

ALL NEW DAZZLING DUNKS AND BASKETBALL BLOOPERS. Spectacular shots and fabulous flubs. (CBS) 48 min., $14.98.

BASKETBALL LAUGHS, GAFFES, AND GOOFS. Bloopers and funny stories taken from the last 40 years of pro, college, and high school hoops. 30 min., $9.98.

DAZZLING DUNKS AND BASKETBALL BLOOPERS. Featuring past and present NBA stars. (CBS) 48 min., $14.98.

FANTASTIC BASKETBALL BLOOPERS. Sold as Doubleheader with BASKETBALL'S AMAZING RAMS, SLAMS, AND JAMS. 30 min. each. $9.95.

HOOPS BLOOPS. Various and sundry spectacular plays gone wrong. (SIM) 30 min., $9.95.

PRO BASKETBALL'S FUNNIEST PRANKS. Hosted by Charles Barkley and Rick Mahorn. 30 min., $9.98.

FOOTBALL

Whenever fans start talking about sports videos, they inevitably get around to NFL Films, and particularly the top-selling series of follies and funnies. The company, the series, and eventually all sports videos got a boost back in 1969 when excerpts from *Football Follies* were shown on the Johnny Carson show. They were funny then and they still hold up today.

Although the funny videos are their best-known efforts, NFL Films is anything but a one-trick pony. Sometimes it seems like ESPN would be blank half the time without NFL Films. Among their offerings are team highlights, highlights from every Super Bowl, football history videos, and such oddities as *Boom! Bang! Whap! Doink! John Madden on Football.*

One of the secrets of NFL Films' success is that they are so complete in their coverage of pro football. What you see on one of their videos is not just a rehash of the same network highlight shots shown on the late news. The company sends its own photography crew to every NFL game to get the whole story. Everything is filmed rather than videotaped because that adds a measure of depth. The company's files contain reels of film detailing every play of every game played in the NFL since the mid-1960s, additional game films they've acquired that go back as far as the 1920s, interviews with just about any pro footballer you can name, and enough sideline and pregame shots to give the viewer that "being there" feeling. No doubt, they could create a feature-length presentation of just "Hi, Mom's." When NFL Films starts putting one of their little programs together, they have an enormous amount of material to choose from.

And they do a good job of melding action with fact. Their writers and editors know the game. Not only can they explain what happens on a given play in Madden-esque detail, but they also have a good grasp on the entire history of the game to put things in perspective. At least in *historical* perspective. If occasionally they make a pro football game seem a bit more important than it really is—say, on the level of a world war combined with the Great Flood—most fans will forgive them. In fact, some of the fans we've seen probably would accuse the company of underplaying.

The voice of NFL Films used to be the late John Facenda, a *presence* who could make an ordinary off-tackle smash resonate as if it had just been delivered down the mountain carved in stone. None of the present narrators are quite up to Facenda's Voice of God, but they still have a knack for hyperbole. Adding to the impact is ever-present dramatic music, in your face closeups, and a sometimes maddening reliance upon slow motion. It's all designed to make everything bigger than life.

Not every football video is put out by NFL Films, of course. For example, CBS/Fox has a worthwhile effort in *Monday Night Madness: The Very Best of Monday Night Football*. And most of the instructionals and college highlight videos come from other sources. However, NFL Films does dominate the market.

Team Highlights

Season highlight videos, which are available for some teams for the past ten years, have the same basic plot structures you'll find in other sports. A video of a successful team stresses its division championsship (*1989 Minnesota Vikings: Division Champions ... Again*), conference title (*1987 Denver Broncos: Champions Against All Odds*), or Super Bowl victory (*World Champions! The Story of the 1985 Chicago Bears*), depending on how far the team progressed in the league's annual bloodletting. If a team had a terrible year, like, say 4–11, the video emphasizes building for the future (*1987 Kansas City Chiefs: Keeping the Faith*). Even 1–15 can be sold as how a team "prepared themselves for a climb back to the top of the NFL," to quote from the NFL Films catalog entry for *1989 Dallas Cowboys: Back to the Future*.

You might like to follow the spin doctors through the good times of the Tampa Bay Buccaneers from 1986 when they went 2–14

(*Countdown to '87*), through 1987 when they soared to 4–11 (*The Start of Something Big*), on to 1988 when they smashed their way to 5–11 (*Team on the Rise*), to 1989 when they coagulated at 5–11 again (*We're Buccaneers*). "Success for the 1989 Buccaneers was measured in their ability to come from behind and fight to the finish," according to the catalog. It's possible Tampa Bay fans would have preferred measuring in victories, but NFL Films goes with what it has, and sometimes the task is to make chicken salad out of chicken feathers.

If you're one of those grumps who insist that today's game doesn't rank with the one you grew up with, you might prefer peering at one of the "Special Team Highlight" videos that cover the glory seasons of the best teams of the 1960s and 1970s. *Years to Remember: The New York Giants 1958, 1959, 1961, 1962, 1963* is more than two hours of Sam Huff, Andy Robustelli, Frank Gifford, Y. A. Tittle and the rest of the bridesmaid crew of those seasons. *The Legacy Begins: The Miami Dolphins 1970–74* is highlighted by the NFL's only perfect season to date. Also available are videos on the Packers' and Steelers' Super Bowl teams and several clubs that were only a step below like Cleveland in the the late 1960s. The Dallas Cowboys are available on three separate videos: *The Star Ascending: 1965–69*, *Coming of Age: 1970–74*, and *America's Team: 1975–79*.

Collegiate highlight videos are not as common as their NFL brothers, but NFL Films can sell you *Field of Honor: 100 Years of Army Football* which will give you Blanchard and Davis along with some rare archival footage and a lot of "Beat Navy!"

For History Buffs

The Special Team Highlights and, for that matter, the yearly team highlights are, in effect, history videos. However, to get a more complete story on your favorite team, NFL Films puts out several team histories. Although the span covered is longer than in the Special Team Highlights, a lot of detail gets lost. For example, *Giants Forever: A History of the New York Giants* covers the team from 1925 yet is 95 minutes shorter than the video on the early 1960s. *Shoot for the Stars*, billed as "a concise history of the Cowboys from their days as a ragtag 1960 expansion team," runs 40 minutes; the three videos covering 1965 to 1979 total 366 minutes.

In other words, the team histories are the *Reader's Digest* versions.

Nevertheless, you can get a pretty complete reading of the entire history of pro football through videos. You might start with *The History of Pro Football*, which covers the span from Red Grange to the Super Bowl in 87 minutes. Or you might prefer *Big Game America*, with Burt Lancaster narrating the events of the NFL's first 50 years. Then you can peruse *The Fabulous Fifties* (two hour-long volumes), *Sensational '60s*, *The Super Seventies*, and *Era of Excellence—The 1980s*.

For college fans, *Greatest Moments in College Football* has a lot to offer, but could probably double its 48 minutes.

Individual Games

If you're a Super Bowl junkie, you can get an individual video of every one of the Roman-numeraled orgies, but, let's face it, most of the games have been dogs. The videos generally run about 23 minutes and there have been games that didn't have that much worthwhile action. You might be happier with some of the *Highlights of Super Bowl* collections. The cost of these is about twice that of a single Super Bowl video, but each runs approximately 90 minutes and gives you four games.

However, for exciting football action, you can't beat *NFL's Greatest Games, Volumes 1 & 2* (also available separately) with highlights from a dozen of the best contests ever. It includes the Colts' overtime win against the Giants for the 1958 Championship, often called (mistakenly, I think) "the greatest game ever played." Other, and possibly better, games are the Browns' last-second win over the Rams in the 1950 Championship Game, Green Bay's "Ice Bowl" victory over Dallas in 1967, and the Chargers' super-overtime playoff triumph against the Dolphins in 1982.

Like anything else in sports, a lot of personal feelings go into picking "the greatest" whatever. A Packers fan might opt for the Ice Bowl or any one of a dozen other Green Bay games of the 1960s. We're old enough to remember that Browns win in 1950, and the sight of Otto Graham faultlessly running what we only later called a "two-minute" drill nearly brought tears to my eyes.

There are also three *Most Memorable Games of the Decade* videos available on the 1970s.

It's only fair to say that there is a certain amount of repetition involved here. Certain big plays pop up on more than one video. Sooner or later, you may get a little tired of watching Bart Starr mush into the endzone on that cold day in 1967 and want to fast-forward to something else. That's the problem with Big Plays in Big Games; they keep cropping up in different story lines. As big as Starr's sneak was, it was still only a quarterback sneak—tremendously important, but not nearly as spectacular as hundreds of Walter Payton dashes that highlighted otherwise unmemorable Bears games. If you want to see some run by Jim Brown or Gale Sayers or O. J. that you remember as "The Greatest Ever," you may have to check out quite a few videos before you find it. All the same, if it really was a sensational scamper you can rest assured that NFL Films has it tucked into one of its offerings.

Probably the first place to look for your favorite play should be in one of *The NFL's Best Ever* series. The last time I checked, they were offering "best ever" runners, quarterbacks, coaches, teams, and professionals.

Individual Stars

NFL Films doesn't regularly devote an entire video to one player no matter how much of a superstar he may be. The only exception seems to be *Lombardi*, a full 51 minutes on the Packers' charismatic coach of the 1960s. Generally, though, individuals are profiled a half dozen or so at a time in special programs like *Gift of Grab* (receivers), *Heart of a Champion* (courageous players), *Tough Guys* (both mentally and physically), and *NFL Crunch Course* (defenders past and present).

The good news is that you see the guy in action, hear a little of his philosophy, and move on before he becomes boring. The bad news is that you almost never learn enough to make the guy seem an interesting personality. Assuming that at least some of them are worth knowing more about, this is a weakness.

Those Funny Videos

According to the head of "Sports Books, etc.," funny baseball videos outsell the football yukkers at his emporium except around Christmastime. However, he also points out that baseball *anything* outsells any other sport among his customers. Actual total figures are impossible to come by, but I *think*—and don't make me swear to it—that football humor has sold more videos overall than baseball. They've been around longer, and also have been "bundled" with more magazine deals.

Not everyone may agree with this, but I can sit through a supposedly humorous baseball video and only occasionally crack a smile and yet get several—sometimes many—audible chuckles out of one of NFL Films' funnies. Since I like baseball just as much as I like pro football, I wondered where the difference lay. To that end, I have propounded several theories.

Theory #1: *Cartoons vs. People.* You may have noticed that a generic fat man slipping on a banana peel is funnier than when your grandmother flops on the same discarded peel. Obviously, part of the difference is that the fat man is a nameless cartoon-like presence that allows us to concentrate on the action (i.e., feet in the air, rump to the floor). When Grandma slips we worry about her hip.

Something of the same thing happens in our baseball-football humorous video. The footballers, with their Darth Vader helmets and Li'l Abner shoulders, don't look like real people out there. They belong with Syvester and Tweety. But baseballers in their uniforms still maintain recognizably human shapes. So we laugh at a discombobulated football player who is an alien presence and worry about a disencumbered baseball player who is like us. Adding to our discomfort is the knowledge that footballers always play with a load of pain whereas a hangnail can put a baseballer on the DL for six weeks.

Theory #2: *Handsome Is as Handsome Does.* Put a baseballer player on the bench and have him do something that's supposed to be funny like squirting shaving cream atop the hat of a teammate, and you see a nice-looking young man pulling a sophomoric prank. The word "smartass" is certain to inhabit one corner of your mind. When the bench-sitting footballer turns to the camera to say "Hi, Mom," you usually see a lop-sided, gap-toothed, nose-broken grin that only Mom could love. Do we hear the word "clown"?

Theory #3: *It's All in the Ball.* A baseball is round. In most cases, it bounces just about where you'd expect it to bounce. A football is a little blimp with pointy ends. Nothing else worthwhile is shaped like that thing, and nothing else bounces the way—or every which way—it does. When a shortstop juggles a pop fly, it may bounce once, twice, even three times, but if he catches it, the play is over and if he doesn't it becomes a different kind of play. When a football is fumbled, it can go anywhere, be touched by many different hands, and having landed continues to be the object of a mad scramble.

Theory #4: *Production Values.* NFL Films consistently weds goofy sound-effects and clever music cuts to its humor. Our absolute favorite is the segment in which footballers along the sideline apparently sing "Hi, Mom; we're number one!" to *Largo Al Factoum* from *The Barber of Seville* by Rossini. Baseball videos have attempted to do some of these things, but the Boing! of an outfielder bouncing off a wall just isn't as funny as the Crrr-UNCH! of a ball carrier being buried beneath a thousand blubbery pounds of defensive linemen.

Theory #5: *The Nature of the Beast.* Despite Herculean efforts on the parts of some players, umpires, league officials, and agents, baseball is still a fun game that can be played for laughs in your own backyard. As grimly as major league baseball is played, we go to a baseball park to have fun, and if our team loses, there's always tomorrow and the sun will shine. A moderately amusing happenstance is just another chuckle and can get lost in a sea of amusement.

On the other hand, only a few real wars have been enacted with the ponderous, tight-lipped seriousness of an NFL game, especially as reconstructed in one of NFL Films' "straight" videos. When we see pro football actually laughing at itself, we can't resist joining in. It's as if Moses had come down from the mountain and revealed the first commandment to be "Take my wife—please!"

Theory #6: *The Fustest with the Bestest.* This is the most basic theory of all. It simply suggests NFL Films started this kind of video and has stayed consistently ahead of its competition. One reason I lean toward this theory is the presence of *Pro Football Funnies* among our collection. At first glance, you'd figure this to be another one in NFL Films series. But on closer examination, it comes from Halcyon Days Productions and the game situations are from the late and sometimes lamented United States Football League. The video tries hard, but it just doesn't pack the giggles of the NFL Films jobs. One handicap, of course, is that the USFL lasted only three seasons, thus giving Halcyon considerably less footage to choose from. Nevertheless, everything done here has been done better by the big guys.

All of this is not to say that every frame of an NFL Films production is hilarious. Sometimes they reach too far. Forcing pro football film into an old movie format (*NFL Follies Goes Hollywood*) or a TV show burlesque (*The NFL TV Follies*) strikes us as funnier in concept than in execution. If you're going to buy only one Follies video, we'd suggest you go with one of the highlight videos that put together many of the best pieces from other Follies videos—either *Best of Football Follies* or *Football Follies on Parade*.

Instructionals

NFL Films doesn't put out instructionals, unless you want to count *Boom! Bang! Whap! Doink! John Madden on Football,* which is really about how to watch a game. *NFL Playbook—A Fan's Guide to Flea Flickers, Fumbles and Fly Patterns* sounds like an X's and O's video but is really only a compilation of plays, some spectacular and some odd.

One of the reasons football instructionals don't sell as well as baseball instructionals do is that younger kids, the junior high and grade school set, are more likely to be out playing soccer, which is less likely to break those developing bones. As the kid gets into high school and decides he'd like to be Mr. Touchdown, he's already got several fulltime coaches. What's he need to study a video for?

It's also true that a bunch of kids sitting around on a Saturday morning can play soccer if they've got a ball and a field, basketball if they've got a ball, hoop, and driveway or baseball if they can find a ball, bat, and field. But not too many are going to opt for tackle football on a rock-strewn field unless they have helmets and shoulder pads. In other words, the amount of expensive equipment required to play football mitigates against many kids studying skills on their own.

The best market for a football instructional is a football coach.

Although football instructionals are not as popular as baseball instructionals, a few are available. As with other sports, you want to gear the video to the level of the kids receiving instruction. For beginners, *The Official Pop Warner Video Handbook* is a basic instructional guide. *Sports Clinic: Football* with George Allen and others is a little more advanced but still deals with basics of line play, receiving and quarterbacking. And, if you figure your kid as another Joe Montana, you can give him *Quarterbacking Fundamentals & Techniques* with Maryland coach Joe Krivak and two NFL backups he coached, Frank Reich and Stan Gelbaugh.

Football Videos

College Football: General Highlights

Traditionally, college football precedes pro ball in any compendium. The distinction is neither alphabetical nor chronological; years ago pro football was a minor sport compared to the college game, and, as such, was relegated to the back pages of preseason magazines. The NFL has grown up, but we'll stick with the old fashion.

BIG TEN COLLEGE FOOTBALL HIGHLIGHTS 1958 AND 1971. Two seasons of highlights from one of the most respected conferences. (RSF) 57 min., $29.95.

FOOTBALL CLASSICS, VOLS. I and II. Two tapes containing great moments and features from both college and pro football from the 1920s through the 1960s. (VES) 60 min., $14.95 each.

GREAT MOMENTS IN COLLEGE FOOTBALL. Curt Gowdy narrates a history of college football. (1988) 48 min., $29.98.

HERITAGE OF THE HEISMAN TROPHY. The college players voted best of the year, Staubach, Simpson, Plunkett, Dorsett, Flutie, Sanders. 60 min., $19.95.

ROSE BOWL HIGHLIGHTS THROUGH THE YEARS. History of the "Grand-daddy of them all." 60 min., $19.95.

ROSE PARADE THROUGH THE YEARS. For those who prefer floats to footballs. 60 min., $19.95.

College Football: Particular Teams

Any college or university from Notre Dame down to Podunk Tech is capable of producing its own highlight videos. The following are some that have been made available nationally. If your school isn't represented, I suggest you write or call the school's Sports Information Director.

(Alabama) A NEW DECADE, A NEW BEGINNING. Highlights of the Tide's 1990 season. 30 min., $19.99.

(Alabama) 16 to 7: ALABAMA'S AMAZING 1990 IRON BOWL VICTORY OVER AUBURN. 35 min., $14.95.

(Army) FIELD OF HONOR: *100 Years of Army Football.* NFL Films traces football at West Point. Blanchard-Davis, Coach Blaik, etc.— includes rare archival footage. (NFL, 1990) 45 min., $19.98.

AUBURN: *The Decade of the Eighties.* Story of the greatest decade in Auburn football history. 60 min., $29.99.

AUBURN FOOTBALL HIGHLIGHTS. The 1990 season. 40 min., $19.99.

(Florida State) BOBBY BOWDEN: *Building a Tradition.* Fifteen years of Florida State gridiron success. 90 min., $19.99.

(Florida State) GIGGED AGAIN: *The 1990 Florida State Win over Rival Florida.* 50 min., $14.99.

(Florida State) 'NOLES REVIEW: *Florida State Seminoles 1990 Season Highlights.* 75 min., $19.99.

(Georgia) 25 YEARS OF GEORGIA FOOTBALL: *The Vince Dooley Era.* Highlights from Dooley's 201 victories, 1964–88. 100 min., $29.99.

(Michigan) VINTAGE BO. The public and private Bo Schembechler during his second season at Michigan. (1989) 30 min., $19.95.

(Michigan) WOODY VS. BO: *The Ten Year War.* Highlights of Michigan vs. Ohio State 1969–78. 51 min., $19.95.

NOTRE DAME HEISMANS: *The Men and the Moments.* Film highlights and reminiscences by Bertelli, Lujack, Hart, Lattner, Hornung, Huarte, and Brown. (1989) 58 min., $19.99.

(Notre Dame) KNUTE ROCKNE AND THE FIGHTING IRISH. Sixty years after his death in a plane crash, Rockne is still the most legendary football coach of all time; hosted by Paul Hornung. (FRI, 1989) 45 min., $19.95.

(Notre Dame) THEY WANTED TO WIN: *1988 Notre Dame National Championship Season Highlights.* Includes wins over Michigan, Miami, USC, and West Virginia in the Fiesta Bowl. 60 min., $19.99.

NOTRE DAME FOOTBALL 1989. Season highlights, including Orange Bowl win over Colorado. $19.99.

(Ohio State) WOODY VS. BO: *The Ten Year War.* Highlights of Michigan vs. Ohio State 1969–78. 51 min., $19.95.

(Tennessee) ONE HUNDRED YEARS OF VOLUNTEERS. History of Tennessee football, including highlights and interviews with players and coaches. 166 min., 2 tapes, $49.99.

(Tennessee) ONE HUNDRED YEARS OF VOLUNTEERS GALA CELEBRATION. Tape of the Centennial banquet, featuring Dave Loggins' live performance of "Orange Memories." 141 min., $19.99.

(Tennessee) NOVEMBER 10, 1979: *Tennessee 40, Notre Dame 18.* 48 min., $29.99.

(Tennessee) SUGAR VOLS: *Sweet Taste of Sugar* and *Sugar Bowl Memories.* Tennessee's 1985 SEC Championship season highlights, and highlights of the Vols' 35–7 win over Miami in New Orleans. 40 min. and 45 min., $39.99.

(Tennessee) VOLS '87: *The 4th Quarter Belongs to Us.* Highlights of the season and Peach Bowl victory over Indiana. 60 min., $29.99.

(Tennessee) VOLS 1989: *Runnin' in High Cotton*. SEC Championship season and Cotton Bowl victory over Arkansas. 58 min., $29.99.

(Tennessee) SEPTEMBER 9, 1989: *Tennessee 24, UCLA 6*. 80 min., $29.99.

(Tennessee) COTTON BOWL 1990: *A BigPlay in Big D*. 31–27 win over Arkansas caps an 11-1 1990 season. 48 min., $19.99.

(Tennessee) VOLS 1990: *Back on Bourbon Street*. Highlights of centennial season and Sugar Bowl win over Virginia. 63 min., $19.99.

(Texas) WHATEVER IT TAKES: *The 1990 Texas Longhorns SWC Championship Football Season*. $19.99.

Super Bowl Highlights

NFL Films standard Super Bowl highlight video runs 23 minutes and usually retails for $14.95 each. If you're anything like us, you get those Roman numerals mixed up, so we've included the game score and the season.

I. Green Bay 35, Kansas City 10 (1966)

II. Green Bay 33, Oakland 14 (1967)

III. New York Jets 16, Baltimore 7 (1968)

IV. Kansas City 23, Minnesota 7 (1969)

V. Baltimore 16, Dallas 13 (1970)

VI. Dallas 24, Miami 3 (1971)

VII. Miami 14, Washington 7 (1972)

VIII. Miami 24, Minnesota 7 (1973)

IX.   Pittsburgh 16, Minnesota 6 (1974)

X.   Pittsburgh 21, Dallas 17 (1975)

XI.   Oakland 32, Minnesota 14 (1976)

XII.   Dallas 27, Denver 10 (1977)

XIII.   Pittsburgh 35, Dallas 31 (1978)

XIV.   Pittsburgh 31, Los Angeles 19 (1979)

XV.   Oakland 27, Philadelphia 10 (1980)

XVI.   San Francisco 26, Cincinnati 21 (1981)

XVII.   Washington 27, Miami 17 (1982)

XVIII. Los Angeles Raiders 38, Washington 9 (1983)

XIX.   San Francisco 38, Miami 16 (1984)

XX.   Chicago 46, New England 10 (1985)

XXI.   New York Giants 39, Denver Broncos 20 (1986)

XXII.   Washington 42, Denver 10. Includes 1987–88 Gillette MVP. (1987) 44 min.

XXIII. San Francisco 20, Cincinnati 16. Includes 1988–89 Gillette MVP. (1988) 44 min.

XXIV.   San Francisco 55, Denver 10. ("CORONATION") (1989)

XXV.   New York Giants 20, Buffalo 19. (1990)

XXVI.   Washington 37, Buffalo 24. (1991)

Super Bowl highlights are also available in four-bowl chunks from NFL Films, with each video running approximately 90 minutes. Suggested retail price is $29.98. The suggested price for the entire collection of six videos is $119.98.

HIGHLIGHTS OF SUPER BOWL I–IV, VOLUME 1

HIGHLIGHTS OF SUPER BOWL V–VIII, VOLUME 2

HIGHLIGHTS OF SUPER BOWL IX–XII, VOLUME 3

HIGHLIGHTS OF SUPER BOWL XIII–XVI, VOLUME 4

HIGHLIGHTS OF SUPER BOWL XVII–XX, VOLUME 5

HIGHLIGHTS OF SUPER BOWL XXI–XIV, VOLUME 6

And finally, for those who want the best in one quick sitting:

MOST MEMORABLE MOMENTS IN SUPER BOWL HISTORY. Highlights from the good, the bad, the happy, and the sad. (NFL, 1990) 50 min., $19.98.

SUPER SUNDAY: *A History of the Super Bowl.* The quick tour from I to XXIV. (NFL, 1987) 55 min., $19.98.

SUPER BOWL DREAM TEAM. Choosing an all-time all-star team from the first 25 Super Bowls. (NFL, 1992) $14.98.

History

BIG GAME AMERICA. Narrated by Burt Lancaster. Reviews the first 50 years of NFL action. (NFL) 51 min., $19.98.

ERA OF EXCELLENCE, THE 1980s. The most exciting players, plays, and games of the decade. (NFL, 1989) 50 min., $19.98.

THE FABULOUS FIFTIES, VOL. I. Frank Gifford, Hardy Brown, Art Donovan and some of the other fabulous personalities of the decade. (NFL) 60 min., $14.98.

THE FABULOUS FIFTIES, VOL. II. The second part of an historic view of the era that has been called "The Golden Age of Pro Football." (NFL) 60 min., $14.98.

THE HISTORY OF PRO FOOTBALL. Review of the game's history

from the 1920s and Red Grange through the Super Bowls. An excellent primer. (NFL) 87 min., $19.98.

MONDAY NIGHT MADNESS. Compilation of the best moments from ABC's popular *Monday Night Football*. (1989) 50 min., $19.98.

NFL '81. Review of the exciting 1981 NFL season; includes the sensational 41–38 San Diego-Miami overtime playoff game. (NFL) 47 min., $14.98.

NFL '87. Review of the 1987 season, highlighted by the emergence of the Redskins and the Cinderella Saints. (NFL, 1988) 47 min., $14.98.

SAVIORS, SAINTS AND SINNERS. Action from the 1980 season, including the playoffs and Super Bowl, plus NFL Films' all-NFL team and Coach of the Year. (NFL) 50 min., $14.98.

SENSATIONAL '60S. The era of Vince Lombardi and the upstart American Football League climaxed by the Jets' upset Super Bowl victory. (NFL, 1988) 60 min., $14.98.

SILVER CELEBRATION. 25 years of NFL Films highlights, bloopers, best players, greatest runs, etc. (NFL) 60 min., $19.98.

THE SUPER SEVENTIES. George Allen, Broncomania, and the top three runners of the 1970s. (NFL) 49 min., $19.98.

History: Teams

NFL Films' standard team highlight video runs 22–23 minutes with a suggested retail price of $14.98. Highlights for many teams are available for the 1981–85 period. Each 1984 team video (except for San Francisco) is combined with "The Road to Super Bowl XIX," a 47-minute review of the 1984 season. San Francisco's 1984 video is combined with Super Bowl highlights. Videos for the strike-shortened 1982 season were combined with other videos to make a 45-minute program. Videos for the 1983 season are combined with "NFL '83," a season review.

Titles are as follows (1990 and 1991 titles and prices unavailable at press time):

Atlanta Falcons

1986 ATLANTA FALCONS: *Falcon Fever.* (NFL) 23 min. $14.98.

1987 ATLANTA FALCONS: *A Preview of the '88 Falcons.* (NFL) 22 min., $14.98.

1988 ATLANTA FALCONS: *Fighting Falcons.* (NFL) 22 min., $14.98.

1989 ATLANTA FALCONS: *Special Moments—25 Yeras of Falcon Football.* (NFL) 22 min., $14.98.

1990 ATLANTA FALCONS. Highlights. (NFL)

1991 ATLANTA FALCONS. Highlights. (NFL)

Buffalo Bills

YEARS OF GLORY... YEARS OF PAIN: *The 25-Year History of the Buffalo Bills.* The AFL Championships, O. J. Simpson's record-setting years, the playoff teams of the early 1980s. (NFL) 28 min., $19.98.

1986 BUFFALO BILLS: *Good Enough to Dream.* (NFL) 23 min., $14.98.

1987 BUFFALO BILLS: *Something to Shout About.* (NFL) 22 min., $14.98.

1988 BUFFALO BILLS: *A Team, a Town, a Dream.* (NFL) 22 min., $14.98.

1989 BUFFALO BILLS. Highlights. (NFL) 22 min., $14.98.

1990 BUFFALO BILLS. Highlights include Super Bowl XXV. (NFL)

1991 BUFFALO BILLS. Highlights include Super Bowl XXVI. (NFL)

Chicago Bears

1979 CHICAGO BEARS: *Go Bears!*

1984 CHICAGO BEARS: *Fight to the Finish/Road to XIX.* (NFL) 70 min., $14.98.

WORLD CHAMPIONS! STORY OF THE 1985 CHICAGO BEARS. (NFL) 58 min., $19.98.

1986 CHICAGO BEARS: *We Will Be Back.* (NFL) 23 min., $14.98.

1987 CHICAGO BEARS: *Bear Down.* (NFL) 22 min., $14.98.

1988 CHICAGO BEARS: *Champions at Heart.* (NFL) 22 min., $14.98.

1989 CHICAGO BEARS: *Wounded Bears.* (NFL) 22 min., $14.98.

1990 CHICAGO BEARS. Highlights. (NFL)

1991 CHICAGO BEARS. Highlights. (NFL)

Cincinnati Bengals

1981 CINCINNATI BENGALS: *Stripes.* (NFL) 23 min., $14.98.

1986 CINCINNATI BENGALS: *Attack, Attack, Attack.* (NFL) 23 min., $14.98.

1987 CINCINNATI BENGALS: *Great Expectations.* (NFL) 22 min., $14.98.

CINCINNATI BENGALS "STARS IN STRIPES" 1988 VIDEO YEARBOOK. Although "Yearbook" includes highlights from the Cincinnati season, it is not the official highlight videos (NFL) 50 min., $19.98.

1988 CINCINNATI BENGALS: *Men on a Mission.* (NFL) 22 min., $14.98.

1989 CINCINNATI BENGALS: *Unfinished Business.* (NFL) 22 min., $14.98.

1990 CINCINNATI BENGALS. Highlights. (NFL)

1991 CINCINNATI BENGALS. Highlights. (NFL)

Cleveland Browns

A WINNING TRADITION: *The Cleveland Browns 1964, 1965, 1967, 1968, 1969.* When five divisional titles and one World Championship made the Browns a force. (NFL) 137 min., $19.98.

1980 CLEVELAND BROWNS: *Cardiac Kids ... Again.* (NFL) 22 min., $14.98.

1985 CLEVELAND BROWNS: *The Division Winners.* (NFL) 23 min., $14.98.

1986 CLEVELAND BROWNS: *Pandemonium Palace.* (NFL) 22 min., $14.98.

1987 CLEVELAND BROWNS: *No Apologies Necessary.* (NFL) 22 min., $14.98.

1988 CLEVELAND BROWNS: *Strange Season.* (NFL) 22 min., $14.98.

1989 CLEVELAND BROWNS: *Division Champions.* (NFL) 22 min., $14.98.

1990 CLEVELAND BROWNS. Highlights. (NFL)

1991 CLEVELAND BROWNS. Highlights. (NFL)

Dallas Cowboys

SHOOT FOR THE STARS. A concise history of the Dallas Cowboys from their beginning in 1960 through the 1980s. (NFL) 40 min., $14.98.

THE GREATEST MOMENTS IN DALLAS COWBOY HISTORY. As selected by readers of the Dallas *Times Herald.* (NFL) 60 min., $19.98.

THE STAR ASCENDING: *The Dallas Cowboys 1965–69.* The Cowboys won 70 percent of their games with Meredith, Hayes, Howley, and Perkins. (NFL) 124 min., $19.98.

COMING OF AGE: *The Dallas Cowboys 1970–74.* Compilation of five Cowboys' annual highlights, including their first NFC Championship and World Championship seasons. (NFL) 124 min., $19.98.

AMERICA'S TEAM: *The Dallas Cowboys 1975–79.* Highlights of five seasons in which the Cowboys made three Super Bowl appearances. (NFL) 118 min., $19.98.

1980 DALLAS COWBOYS: *Like a Mighty River.* (NFL) 22 min., $14.98.

1981 DALLAS COWBOYS: *Star-Spangled Cowboys.* (NFL) 23 min., $14.98.

1982 DALLAS COWBOYS: *Great Expectations/The Man in the Funny Hat.* (NFL) 45 min., $14.98.

1984 DALLAS COWBOYS: *Silver Season/Road to XIX.* (NFL) 70 min., $14.98.

1985 DALLAS COWBOYS: *The Winning of the East.* (NFL) 23 min., $14.98.

1986 DALLAS COWBOYS: *Make Way for Tomorrow.* (NFL) 22 min., $14.98.

1987 DALLAS COWBOYS: *Blueprint for Victory.* (NFL) 22 min., $14.98.

1988 DALLAS COWBOYS: *A New Day in Dallas.* (NFL) 22 min., $14.98.

1989 DALLAS COWBOYS: *Back to the Future.* (NFL) 22 min., $14.98.

1990 DALLAS COWBOYS. Highlights. (NFL)

1991 DALLAS COWBOYS. Highlights. (NFL)

Denver Broncos

BRONCOMANIA. A look at Denver's love affair with the Broncos. (NFL) 30 Min., $14.98.

FIRST TASTES OF GLORY: *The Denver Broncos 1977, 1978, 1979, 1984.* Highlights from four of the Broncos' most outstanding seasons. (NFL) 93 min., $19.95.

1983 DENVER BRONCOS: *A Team Together/NFL '83.* (NFL) 46 min., $14.98.

1984 DENVER BRONCOS: *The Winning of the West/Road to XIX.* (NFL) 70 min., $14.98.

1986 DENVER BRONCOS: *Mile High Champions.* (NFL) 58 min., $19.98.

DENVER BRONCOS "ROCKY MOUNTAIN MAGIC" 1987 VIDEO YEARBOOK. Although "Yearbook" includes highlights from the Denver seasons, it is not the official highlight videos. (NFL) 50 min., $19.98.

1987 DENVER BRONCOS: *Champions Against All Odds.* (NFL) 22 min., $14.98.

1988 DENVER BRONCOS: *A Season in Review.* (NFL) 22 min., $14.98.

DENVER BRONCOS 1989 VIDEO YEARBOOK: *Team Terrific.* Although "Yearbook" includes highlights from the Denver season, it is not the official highlight video. (NFL) 50 min., $19.98.

1989 DENVER BRONCOS: *Mile High Champions.* (NFL) 45 min., $14.98.

1990 DENVER BRONCOS. Highlights. (NFL)

1991 DENVER BRONCOS. Highlights. (NFL)

Detroit Lions

1983 DETROIT LIONS: *Comeback Champions/NFL '83.* (NFL) 46 min., $14.98.

1986 DETROIT LIONS: *Close Encounters.* (NFL) 23 min., $14.98.

1987 DETROIT LIONS: *Working Our Way Back.* (NFL) 22 min., $14.98.

1988 DETROIT LIONS: *Restore the Roar.* (NFL) 22 min., $14.98.

1989 DETROIT LIONS: *Foundation for the Future.* (NFL) 22 min., $14.98.

1990 DETROIT LIONS. Highlights. (NFL)

1990 DETROIT LIONS. Highlights. (NFL)

Green Bay Packers

DISTANT REPLAY: *Champions on and off the Field.* Story of the 1966 World Champion 1966 Packers. 75 min.

THREE IN A ROW: *The Green Bay Packers 1965, 1966, 1967.* The legendary Packers won three championships with Starr, Hornung, Taylor, Nitschke, Adderley, etc. (NFL) 119 min., $19.98.

1986 GREEN BAY PACKERS: *A New Beginning.* (NFL) 23 min., $14.98.

1987 GREEN BAY PACKERS: *Pack to the Future.* (NFL) 22 min., $14.98.

1988 GREEN BAY PACKERS: *Coming Together.* (NFL) 22 min., $14.98.

1989 GREEN BAY PACKERS: *Out of the Pack.* (NFL) 22 min., $14.98.

1990 GREEN BAY PACKERS. Highlights. (NFL)

1991 GREEN BAY PACKERS. Highlights. (NFL)

Houston Oilers

1979 HOUSTON OILERS: *Luv Ya, Blue.* (NFL) 22 min., $14.98.

1986 HOUSTON OILERS: *Coming of Age.* (NFL) 23 min., $14.98.

1987 HOUSTON OILERS: *Finding the Way to Win.* (NFL) 22 min., $14.98.

1988 HOUSTON OILERS: *Runnin', Gunnin' Excitement.* (NFL) 22 min., $14.98.

1989 HOUSTON OILERS: *Grasping for Greatness.* (NFL) 22 min., $14.98.

1990 HOUSTON OILERS. Highlights. (NFL)

1991 HOUSTON OILERS. Highlights. (NFL)

Indianapolis Colts
(also Baltimore)

GLORY DAYS OF YESTERYEAR: *The Baltimore Colts 1964, 1965, 1968, 1970.* Four of the Colts' best seasons, including Tom Matte's session as an "instant quarterback." (NFL) 110 min., $19.98.

1986 INDIANAPOLIS COLTS: *A Bold New Spirit.* (NFL) 23 min., $14.98.

1987 INDIANAPOLIS COLTS: *Off and Running.* (NFL) 22 min., $14.98.

1988 INDIANAPOLIS COLTS: *A Test of Character.* (NFL) 22 min., $14.98.

1989 INDIANAPOLIS COLTS: *Hitting Full Stride.* (NFL) 22 min., $14.98.

1990 INDIANAPOLIS COLTS. Highlights. (NFL)

1991 INDIANAPOLIS COLTS. Highlights. (NFL)

Kansas City Chiefs

1986 KANSAS CITY CHIEFS: *Flight to Prominence.* (NFL) 23 min., $14.98.

1987 KANSAS CITY CHIEFS: *Keeping the Faith.* (NFL) 22 min., $14.98.

1988 KANSAS CITY CHIEFS: *Raising the Level of Expectations.* (NFL) 22 min., $14.98.

1989 KANSAS CITY CHIEFS: *Winning Is an Attitude.* (NFL) 22 min., $14.98.

1990 KANSAS CITY CHIEFS. Highlights. (NFL)

1991 KANSAS CITY CHIEFS. Highlights. (NFL)

Los Angeles Raiders
(also Oakland)

1980 OAKLAND RAIDERS: *Our Finest Hour.* (NFL) 22 min., $14.98.

1982 LOS ANGELES RAIDERS: *Commitment to Excellence '82.* (NFL) 46 min., $14.98.

1983 LOS ANGELES RAIDERS: *Just Win, Baby/NFL '83.* (NFL) 46 min., $14.98.

1985 LOS ANGELES RAIDERS: *Year of Glory.* (NFL) 23 min., $14.98.

1987 LOS ANGELES RAIDERS: *Will to Win (1982–87).* (NFL) 22 min., $14.98.

1990 LOS ANGELES RAIDERS. Highlights. (NFL)

1991 LOS ANGELES RAIDERS. Highlights. (NFL)

Los Angeles Rams

1983 LOS ANGELES RAMS: *Return ofthe Rams/NFL '83.* (NFL) 46 min., $14.98.

1984 LOS ANGELES RAMS: *A Family Tradition/Road to XIX.* (NFL) 70 min., $14.98.

1986 LOS ANGELES RAMS: *Armed and Dangerous.* (NFL) 23 min., $14.98.

1987 LOS ANGELES RAMS: *The Promise and the Challenge.* (NFL) 22 min., $14.98.

1988 LOS ANGELES RAMS: *One for the Books.* (NFL) 22 min., $14.98.

1989 LOS ANGELES RAMS: *Fight to the Finish.* (NFL) 22 min., $14.98.

1990 LOS ANGELES RAMS. Highlights. (NFL)

1991 LOS ANGELES RAMS. Highlights. (NFL)

Miami Dolphins

THE LEGACY BEGINS: *The Miami Dolphins 1970–74.* Years worth remembering—three Super Bowls, two wins, and one perfect season. (NFL) 124 min., $19.98.

1981 MIAMI DOLPHINS: *Champions of the East.* (NFL) 23 min., $14.98.

1982 MIAMI DOLPHINS: *Day of the Dolphin/NFL '82.* (NFL) 46 min., $14.98.

1983 MIAMI DOLPHINS: *Day of Frustration, Season of Triumph/ NFL '83*. (NFL) 46 min., $14.98.

1984 MIAMI DOLPHINS: *Movers, Shakers, * Record Breakers/Road to XIX*. (NFL) 70 min., $14.98.

1985 MIAMI DOLPHINS: *Fight to the Finish*. (NFL) 23 min., $14.98.

1986 MIAMI DOLPHINS: *Rollercoaster Season*. (NFL) 23 min., $14.98.

1987 MIAMI DOLPHINS: *Foundation for the Future*. (NFL) 22 min., $14.98.

1988 MIAMI DOLPHINS: *The New Generation*. (NFL) 22 min., $14.98.

1989 MIAMI DOLPHINS: *Prelude to Glory*. (NFL) 22 min., $14.98.

1990 MIAMI DOLPHINS. Highlights. (NFL)

1991 MIAMI DOLPHINS. Highlights. (NFL)

Minnesota Vikings

THE PURPLE POWER YEARS: *The Minnesota Vikings 1969, 1973, 1974, 1976*. The Vikings went to the Super Bowl four times; Kapp, Tarkenton, Foreman. (NFL) 96 min., $19.98.

1986 MINNESOTA VIKINGS: *Back on the Attack*. (NFL) 23 min., $14.98.

1987 MINNESOTA VIKINGS: *Making a Move*. (NFL) 22 min., $14.98.

1988 MINNESOTA VIKINGS: *A Time to Step Forward*. (NFL) 22 min., $14.98.

1989 MINNESOTA VIKINGS: *Division Champs... Again*. (NFL) 22 min., $14.98.

1990 MINNESOTA VIKINGS. Highlights. (NFL)

1991 MINNESOTA VIKINGS. Highlights. (NFL)

New England Patriots

1985 NEW ENGLAND PATRIOTS: *AFC Champions.* (NFL) 23 min., $14.98.

1986 NEW ENGLAND PATRIOTS: *Fight to the Finish.* (NFL) 23 min., $14.98.

1987 NEW ENGLAND PATRIOTS: *Heart of a Champion.* (NFL) 22 min., $14.98.

1988 NEW ENGLAND PATRIOTS: *A Team of Character.* (NFL) 22 min., $14.98.

1989 NEW ENGLAND PATRIOTS: *Striving for Success.* (NFL) 22 min., $14.98.

1990 NEW ENGLAND PATRIOTS. Highlights. (NFL)

1991 NEW ENGLAND PATRIOTS. Highlights. (NFL)

New Orleans Saints

1986 NEW ORLEANS SAINTS: *Destiny in the Dome.* (NFL) 23 min., $14.98.

1987 NEW ORLEANS SAINTS: *Winners.* (NFL) 22 min., $14.98.

1988 NEW ORLEANS SAINTS: *Tew Tradition—Winning.* (NFL) 22 min., $14.98.

1989 NEW ORLEANS SAINTS: *A Fight to the Finish.* (NFL) 22 min., $14.98.

1991 NEW ORLEANS SAINTS. Highlights. (NFL)

New York Giants

GIANTS FOREVER: *A History of the New York Giants.* The history of the franchise from 1925 through the 1986 Super Bowl victory. (NFL, 1988) 45 min., $19.98.

YEARS TO REMEMBER: *The New York Giants 1958, 1959, 1961, 1962, 1963.* Huff, Robustelli, Katcavage and the DEE-fense; Tittle, Gifford, and Webster on the offense! (NFL) 140 min., $19.98.

1981 NEW YORK GIANTS: *A Giant Step.* (NFL) 23 min., $14.98.

1984 NEW YORK GIANTS: *Giants Again/Road to XIX.* (NFL) 70 min., $14.98.

1986 NEW YORK GIANTS: *Giants Among Men.* (NFL) 58 min. $19.98.

1987 NEW YORK GIANTS: *Back to the Future.* (NFL) 22 min., $14.98.

1988 NEW YORK GIANTS: *The Pride Is Back.* (NFL) 22 min., $14.98.

1989 NEW YORK GIANTS: *Giant Achievers.* (NFL) 22 min., $14.98.

1990 NEW YORK GIANTS. Highlights, including Super Bowl XXV. (NFL) 22 min., $14.98.

1991 NEW YORK GIANTS. Highlights. (NFL)

New York Jets

THE WAY WE WERE: *The New York Jets—Their First 25 Years.* From the dreary days of Harry Wismer's Titans through Namath, Maynard, Gastineau. (NFL) 23 min., $19.98.

1981 NEW YORK JETS: *Talk of the Town.* (NFL) 23 min., $14.98.

1982 NEW YORK JETS: *The Road Warriors/NFL '82.* (NFL) 46 min., $14.98.

1986 NEW YORK JETS: *Maximum Effort.* (NFL) 23 min., $14.98.

1987 NEW YORK JETS: *Playing Hard.* (NFL) 22 min., $14.98.

1988 NEW YORK JETS: *An Affirmationof Pride.* (NFL) 22 min., $14.98.

1989 NEW YORK JETS: *A Breath of Fresh Air.* (NFL) 22 min., $14.98.

1990 NEW YORK JETS. Highlights. (NFL)

1991 NEW YORK JETS. Highlights. (NFL)

Philadelphia Eagles

1979 PHILADELPHIA EAGLES: *The Pride of Eagle Football.* (NFL) 22 min., $14.98.

1986 PHILADELPHIA EAGLES: *Coming of Age.* (NFL) 23 min., $14.98.

1987 PHILADELPHIA EAGLES: *Pride in Their Stride.* (NFL) 22 min., $14.98.

1988 PHILADELPHIA EAGLES: *Living on the Edge.* (NFL) 22 min., $14.98.

1989 PHILADELPHIA EAGLES: *One Tough Team.* (NFL) 22 min., $14.98.

1990 PHILADELPHIA EAGLES. Highlights. (NFL)

1991 PHILADELPHIA EAGLES. Highlights. (NFL)

Phoenix Cardinals
(also St. Louis Cardinals)

1986 ST. LOUIS CARDINALS: *The Right Direction.* (NFL) 23 min., $14.98.

1987 ST. LOUIS CARDINALS: *The 1987 St. Louis Cardinals.* (NFL) 22 min., $14.98.

1988 PHOENIX CARDINALS: *Welcome to the NFL, Arizona.* (NFL) 22 min., $14.98.

1989 PHOENIX CARDINALS: *The Eldest Begins Anew.* (NFL) 22 min., $14.98.

1990 PHOENIX CARDINALS. Highlights. (NFL)

1991 PHOENIX CARDINALS. Highlights. (NFL)

Pittsburgh Steelers

THE CHAMPIONSHIP YEARS: *The Pittsburgh Steelers 1975, 1976, 1979, 1980.* After 40 frustrating years, the Steelers won four Super Bowls; Greene, Bradshaw, Harris, Lambert, Ham, etc. (NFL) 96 min., $19.98.

1982 PITTSBURGH STEELERS: *Steel Tough Town/Steelers 50 Seasons.* (NFL) 45 min., $14.98.

1983 PITTSBURGH STEELERS: *The Right Stuff/NFL '83.* (NFL) 46 min., $14.98.

1984 PITTSBURGH STEELERS: *A New Beginning/Road to XIX.* (NFL) 70 min., $14.98.

1986 PITTSBURGH STEELERS: *Tale of Two Seasons.* (NFL) 23 min., $14.98.

1987 PITTSBURGH STEELERS: *Winning Ways.* (NFL) 22 min., $14.98.

1988 PITTSBURGH STEELERS: *Forging the Future.* (NFL) 22 min., $14.98.

1989 PITTSBURGH STEELERS: *Yes, We Can!* (NFL) 22 min., $14.98.

1990 PITTSBURGH STEELERS. Highlights. (NFL)

1991 PITTSBURGH STEELERS. Highlights. (NFL)

San Diego Chargers

LEGEND OF THE LIGHTNING BOLT: *History of the San Diego Chargers*. Gillman, Lincoln, Ladd, Fouts, Winslow. (NFL, 1986) 40 min., $14.98.

1980 SAN DIEGO CHARGERS: *The Power*. (NFL) 22 min., $14.98.

1981 SAN DIEGO CHARGERS: *Cliffhangers, Comebacks, Character*. (NFL) 23 min., $14.98.

1982 SAN DIEGO CHARGERS: *Team of the '80s/NFL '82*. (NFL) 46 min., $14.98.

1986 SAN DIEGO CHARGERS: *Make Way for Tomorrow*. (NFL) 23 min., $14.98.

1987 SAN DIEGO CHARGERS: *The New Generation*. (NFL) 22 min., $14.98.

1988 SAN DIEGO CHARGERS: *Blueprint for Glory*. (NFL) 22 min., $14.98.

1989 SAN DIEGO CHARGERS: *Re-Charged for the '90S*. (NFL) 22 min., $14.98.

1990 SAN DIEGO CHARGERS. Highlights. (NFL)

1991 SAN DIEGO CHARGERS. Highlights. (NFL)

San Francisco 49ers

GALLANT MEN AND GOLDEN MOMENTS: *San Francisco's First 40 Years*. From their beginning in 1946 in the AAFC, spotlighting Tittle, McElhenny, Montana, others. (NFL) 34 min., $19.98.

1981 SAN FRANCISCO 49ERS: *A Very Special Team*. (NFL) 23 min., $14.98.

1983 SAN FRANCISCO 49ERS: *Back Among the Best/NFL '83*. (NFL) 46 min., $14.98.

1984 SAN FRANCISCO 49ERS: *A Team Above All/Super Bowl XIX Highlights*. (NFL) 46 min., $14.98.

1985 SAN FRANCISCO 49ERS: *Never Surrender*. (NFL) 23 min., $14.98.

1986 SAN FRANCISCO 49ERS: *Heart of a Chanpion*. (NFL) 23 min., $14.98.

1987 SAN FRANCISCO 49ERS: *One Heartbeat*. (NFL) 22 min., $14.98.

SAN FRANCISCO 49ERS "TEAM OF THE DECADE" 1988 VIDEO YEARBOOK. Although "Yearbook" includes highlights from the San Francisco season, it is not the official highlight videos. (NFL) 50 min., $19.98.

1988 SAN FRANCISCO 49ERS: *State of the Art*. (NFL) 22 min., $14.98.

SAN FRANCISCO 49ERS "MASTERS OF THE GAME" 1989 VIDEO YEARBOOOK. Although "Yearbook" includes highlights from the San Francisco season, it is not the official highlight video. (NFL) 50 min. $19.98.

1989 SAN FRANCISCO 49ERS: *Back to Back*. (NFL) 22 min., $14.98.

1990 SAN FRANCISCO 49ERS. Highlights. (NFL)

1991 SAN FRANCISCO 49ERS. Highlights. (NFL)

Seattle Seahawks

1983 SEATTLE SEAHAWKS: *The Cinderella Seahawks/NFL '83*. (NFL) 46 min., $14.98.

1984 SEATTLE SEAHAWKS: *One from the Heart/Road to XIX*. (NFL) 70 min., $14.98.

1986 SEATTLE SEAHAWKS: *A Season in Three Acts*. (NFL) 23 min., $14.98.

1987 SEATTLE SEAHAWKS: *Back to the Playoffs.* (NFL) 22 min., $14.98.

1988 SEATTLE SEAHAWKS: *Champions of the West.* (NFL) 22 min., $14.98.

1989 SEATTLE SEAHAWKS: *Northwest Passage.* (NFL) 22 min., $14.98.

1990 SEATTLE SEAHAWKS. Highlights. (NFL)

1991 SEATTLE SEAHAWKS. Highlights. (NFL)

Tampa Bay Buccaneers

1979 TAMPA BAY BUCCANEERS: *From Worst to First.* (NFL) 22 min., $14.98.

1986 TAMPA BAY BUCCANEERS: *Countdown to '87.* (NFL) 23 min., $14.98.

1987 TAMPA BAY BUCCANEERS: *The Start of Something Big.* (NFL) 22 min., $14.98.

1988 TAMPA BAY BUCCANEERS: *Team on the Rise.* (NFL) 22 min., $14.98.

1989 TAMPA BAY BUCCANEERS: *We're Buccaneers.* (NFL) 22 min., $14.98.

1990 TAMPA BAY BUCCANEERS. Highlights. (NFL)

1991 TAMPA BAY BUCCANEERS. Highlights. (NFL)

Washington Redskins

HAIL TO THE REDSKINS: *Their First 50 Years.* Sammy Baugh, George Marshall, Bobby Mitchell, George Allen, Joe Gibbs, and countless others. (NFL) 23 min., $14.98.

THREE CHEERS FOR THE REDSKINS: *The Washington Redskins*

1971. George Allen takes the Redskins from a losing record to the playoffs. (NFL) 53 min., $19.98.

1981 & 1982 WASHINGTON REDSKINS: *Two Years to the Title.* (NFL) 46 min., $14.98.

1983 WASHINGTON REDSKINS: *A Cut Above/NFL '83.* (NFL) 46 min., $14.98.

1984 WASHINGTON REDSKINS: *Winners and Still Champions/Road to XIX.* (NFL) 70 min., $14.98.

1986 WASHINGTON REDSKINS: *The Next Generation.* (NFL) 23 min., $14.98.

WASHINGTON REDSKINS "WARPATH" 1987 VIDEO YEARBOOK. Although "Yearbook" includes highlights from the Denver seasons, it is not the official highlight videos. (NFL) 50 min., $19.98.

1987 WASHINGTON REDSKINS: *Second to None.* (NFL) 22 min., $14.98.

1988 WASHINGTON REDSKINS: *Keeping the Faith.* (NFL) 22 min., $14.98.

1989 WASHINGTON REDSKINS: *Backon the Warpath.* (NFL) 22 min., $14.98.

1990 WASHINGTON REDSKINS. Highlights. (NFL)

1991 WASHINGTON REDSKINS. Highlights, includes Super Bowl XXVI. (NFL)

History and Highlights: Special Programs

Rather than devoting a complete video to an individual player, NFL Films tends to lump eight or ten-minute features on five or six stars together in one video and give it an inclusive title like "Crunchtime." Sometimes the relationships are not clear but the individual segments usually stand up well. While they might be considered as collections of mini-biographies, the overall effect of these videos is to put them in the historic highlight section. Several of these are promotional tapes for various sponsors such as Alcoa, Nestle's Crunch, and Swanson Hungry-Man Dinners. The suggested retail price is $19.98, with a few, such as the "Best Ever"s, going as high as $29.98.

ALCOA'S FANTASTIC FINISHES: *"The Movie."* Includes several examples of last-minute heroics, including Staubach's "Hail Mary" pass and "The Drive" by John Elway vs. Cleveland. (NFL, 1988) 45 min., $19.98.

ALL THE BEST. Untouchable players, unpredictable heroes, and those who battled impossible odds to make the NFL. (NFL, 1987) 60 min., $14.98.

BAD BOYS AND GOOD MEN. Some of the "characters," like Tombstone Jackson and Johnny Blood, profiled. (NFL, 1990) 50 min., $19.98.

BEAUTIES AND THE BEASTS. Mixing profiles of glamour guys like Joe Theismann with "beasts" like Christian Okoye. (NFL, 1990) 50 min., $19.98.

BIG BLOCKS AND KING-SIZE HITS. A treasury of linebackers like Derrick Thomas, Butkus, Lambert, and Taylor. A Hershey Chocolate promo item. (NFL, 1990) 45 min., $19.98.

BIG PLAYS, BEST SHOTS AND BELLY LAUGHS. Some humor but mainly highlights, including Steve Largent setting his receiving records and the coronation of the "Team of the '80s." You have 49 guesses. (NFL, 1990) 50 min., $19.98.

BOMBS AWAY. Some of the top long-ball tossers and receivers in the game's history. (NFL, 1990) 50 min., $19.98.

CRUNCH COURSE. A look at several outstanding defenders, past and present, including Butkus, Deacon Jones, Lawrence Taylor, others. (NFL, 1985) 43 min., $19.98.

CRUNCHTIME. Several tough customers are profiled, including Mike Curtis, Larry Csonka, Howie Long, and Randy White. (NFL, 1986) 44 min., $19.98.

FOOTBALL LEGENDS. Various profiles. (RVI) 40 min. $19.95.

GIFT OF GRAB. Profiling some of the top receivers from the 1940s through the 1980s; Hutson, Berry, Swann, Clark, others . (NFL, 1988) 45 min., $14.98.

THE 1985-86 GILLETTE/NFL MOST VALUABLE PLAYER. Profiles the top six candidates and picks the winner. (NFL) 23 min., $14.98.

THE 1986-87 GILLETTE / NFL MOST VALUABLE PLAYER. (NFL) 23 min., $14.98.

GOLDEN GREATS OF FOOTBALL. Several stars profiled. 45 min., $14.95.

GREATEST MOMENTS OF THE LAST 25 YEARS. Some of the biggest and best achievements on NFL gridirons. (NFL, 1990) 50 min., $19.98.

THE GREAT ONES. Profiles of Jim Brown, Roger Staubach, O.J. Simpson and others in action. (NFL, 1987) 44 min., $19.98.

GREAT TEAMS / GREAT YEARS. Two team highlight films: the 1968 New York Jets who won Super Bowl III and the 1973 Buffalo Bills during the season O. J. Simpson rushed for 2,003 yards. (NFL) 47 min., $19.98.

HEART OF A CHAMPION. Players who have shown exceptional courage, including Doug Williams and Rocky Bleier. (NFL, 1988) 55 min., $19.98.

HIGH STAKES HEROES. Players who had one supreme moment to define their careers: Franco Harris, Joe Namath, Jim O'Brien, Max McGee, and others. (NFL, 1989) 50 min., $19.98.

HIT AFTER HIT. Modest highlights of USFL. 30 min., $9.95.

IN THE CRUNCH. Some of the most memorable, last-minute drives led by Unitas, Bradshaw, Montana and Elway. (NFL, 1987) 60 min., $19.98.

LEGENDARY LINEMEN. Includes Chuck Bednarik, Deacon Jones, Bob Brown, Doug Atkins, and others who excelled in the "trenches." (NFL, 1987) 60 min., $19.98.

MAVERICKS AND MISFITS. Profiles several unique personalities and free spirits of the NFL's past. (NFL, 1987) 60 min., $19.98.

NFL HEAD COACH: *A Self Portrait.* Unnarrated series of candid moments along the NFL sidelines. Please note: this is somewhat dated; few of the men shown are in charge by now. (NFL) 43 min., $29.98.

NFL PLAYBOOK: *A Fan's Guide to Flea Flickers, Fumbles and Fly Patterns.* Although this sounds like an instructional, it's really a highlight. You won't learn how to run a flea flicker, but you may recognize the next one. (NFL, 1989) 50 min., $19.98.

NFL QUARTERBACK. Features Kosar, Elway, Marino and others, along with interviews with teammates and coaches. (NFL, 1988) 45 min., $19.98.

THE NFL'S BEST EVER COACHES. Profiles the careers of legendary leaders like Paul Brown, George Halas, Vince Lombardi, Don Shula, Tom Landry and others. (NFL) 46 min., $29.98.

THE NFL'S BEST EVER PROFESSIONALS. Profiles five legends: Butkus, Billy Kilmer, Larry Brown, Jim Marshall, and Dick Vermeil. (NFL) 46 min., $29.98.

THE NFL'S BEST EVER QUARTERBACKS. The leading quarterbacks in NFL history are analyzed and an all-time best is chosen, along with sensational action footage of great passers at work. (NFL, 1985) 46 min., $29.98.

THE NFL'S BEST EVER RUNNERS. Spectacular running by Jim Brown, O. J. Simpson, Gale Sayers and others. Players analyze their styles. (NFL, 1985) 46 min., $29.98.

THE NFL'S BEST EVER TEAMS. The legendary teams of NFL history, including Lombardi's Packers, Noll's Steelers, Shula's undefeated Dolphins, and others. (NFL, 1981) 46 min., $29.98.

THE NFL'S GREATEST HITS. Not merely an unreeling of big blasts; follows several interesting stories, including some big boo-boos. (NFL, 1989) 50 min., $19.98.

THE NFL'S HUNGRIEST MEN. Highlights of some heroes you might not see elsewhere. A Swanson Hungry-Man promo. (NFL, 1989) 50 min., $19.98.

THE NFL'S HUNGRIEST MEN OF THE '90s. Predicting the players and teams of the 1990s. A Swanson Hungry-Man promo item. (NFL, 1990) 45 min., $19.98.

NFL'S INSPIRATIONAL MEN AND MOMENTS. Three-part video including "Joe and the Magic Bean" (Joe Namath), "Roger Staubach: All-American Hero," and "They Said It Couldn't Be Done" (profiling O. J. Simpson and Fran Tarkenton). (NFL) 52 min., $19.98.

NFL'S ULTIMATE FOOTBALL CHALLENGE. An interactive party tape that includes a trivia quiz.

PLAYING WITH FIRE. Emphasizes several players whose hearts were bigger than their physical assets: Pat Fischer, Joe Kapp. (NFL, 1989) 50 min., $19.98.

SEARCH AND DESTROY. Looks at some of the best defenses and defenders of the last four decades. (NFL, 1989) 50 min., $19.98.

SEE HOW THEY RUN. Some of the NFL's greatest runners; Dorsett, Craig, Campbell, Ricky Bell. (NFL, 1989) 50 min., $19.98.

STRANGE BUT TRUE BODY SHAPES. Hosted by Dan Dierdorf; the tall (Harold Carmichael), small ("Ice Cube" McNeil), the wall ("Refrigerator Perry), and others. (NFL, 1988) 45 min., $19.98.

STRANGE BUT TRUE FOOTBALL STORIES. Hosted by Vincent

Price. Gimmicky retelling of such yarns as the one-eyed quarterback and the player who ate raw meat. (NFL, 1987) 45 min., $19.98.

SUPERSTARS OF THE NFL. Chris Doleman, Keith Millard, Roger Craig, and others from the perspective of 1990. (NFL, 1990) 50 min., $19.98.

THUNDER AND DESTRUCTION. Several of the NFL's hardest hitters are profiled; Lawrence Taylor, Steve Atwater, Bruce Smith, etc. (NFL, 1991) $19.98.

TOUGH GUYS. Hosted by Mike Ditka. Some of the toughest from the 1960s through the 1980s, including Lambert, Fouts, Bavaro and others. (NFL, 1988) 45 min., $19.98.

THE YOUNG, THE OLD AND THE BOLD / TRY AND CATCH THE WIND. Fast-paced profiles of the top passers and receivers of the 1960s, including Namath, Gabriel, Hayes, Warfield, and others. (NFL, 1969) 49 min., $19.98.

History: Individual Games

1958 NFL CHAMPIONSHIP GAME HIGHLIGHTS. Baltimore's "sudden death" victory in overtime has been called "the greatest game ever played"; certainly it's one of the most famous. (NFL) 55 min., 14.98.

1960 NFL CHAMPIONSHIP GAME HIGHLIGHTS. Philadelphia defeated the Packers in the only championship game Lombardi would ever lose. (NFL) 26 min., $14.98.

THE 1966 AND 1967 NFL CHAMPIONSHIP GAMES. The two fabulous victories of the Packers over the Cowboys, both decided in the final seconds. (NFL) 50 min., $14.98.

THE NFL'S GREATEST GAMES. Some of the most heart-stopping contests of NFL history. (NFL, 1986) 44 min., $14.98.

THE NFL'S GREATEST GAMES, VOL. II. More of the NFL's most exciting meetings. (NFL, 1987) 60 min., $14.98.

(Both volumes of The NFL's Greatest Games available on one 104 min. video. $24.98.)

MOST MEMORABLE GAMES OF THE DECADE #1. Highlights from two of the 1970s' biggest games: Dolphins-Chiefs in 1971 and Raiders-Colts in 1977. (NFL) 45 min., $19.98.

MOST MEMORABLE GAMES OF THE DECADE #2. Last-second victories in two classic games from 1974. (NFL) 45 min., $19.98.

MOST MEMORABLE GAMES OF THE DECADE #3. An offensive explosion between the Jets and Colts in 1972; the Raiders earn their way to the Super Bowl by way of a last-second win over New England in the playoffs in 1976. (NFL) 45 min., $19.98.

Individual Heroes

NFL Films has devoted an entire video to only one of pro football's heroes—Vince Lombardi. However, the Greatest Sports Legends series (now Video Cards) presents an inexpensive, though hardly penetrating, view of several players' careers.

LANCE ALWORTH. Video Football Card. (RVI) 30 min., $6.95.

TERRY BRADSHAW. Video Football Card. (RVI) 30 min., $6.95.

FRANK GIFFORD. Video Football Card. (RVI) 30 min., $6.95.

OTTO GRAHAM. Video Football Card. (RVI) 30 min., $6.95.

JOE GREENE. Video Football Card. (RVI) 30 min., $6.95.

PAUL HORNUNG. Video Football Card. (RVI) 30 min., $6.95.

LOMBARDI. NFL Films details the career of the great coach through interviews, filmed practices, and games. (NFL) 51 min., $29.98.

VINCE LOMBARDI. Video Football Card. (RVI) 30 min., $6.95.

WALTER PAYTON. Video Football Card. (RVI) 30 min., $6.95.

GALE SAYERS. Video Football Card. (RVI) 30 min., $6.95.

O.J. SIMPSON. Video Football Card. (RVI) 30 min., $6.95.

BART STARR. Video Football Card. (RVI) 30 min., $6.95.

BART STARR/JOHNNY UNITAS. Doubleheader. (RVI) $14.95.

ROGER STAUBACH. Video Football Card. (RVI) 30 min., $6.95.

JOHNNY UNITAS. Video Football Card. (RVI) 30 min., $6.95.

Instructionals

Football instructionals aren't as common as highlight videos, but there are more than enough for all levels. Among the most intriguing to watch: the six-tape set by the Master, Vince Lombardi, who died in 1970.

BOOM! BANG! WHAP! DOINK! JOHN MADDEN ON FOOTBALL. Madden explains how to get the most out of watching football with classic NFL footage and interviews with Unitas, Namath, Montana, and others. (NFL) 60 min., $19.98.

COACHES VIDEO NETWORK: *ONE-ON-ONE COACHING VIDEOS*. 60 min., $47.50 each.

    DEFENSIVE MIDDLE GUARD. Charlie McBride.

    OFFENSIVE BACKS. Steve Axman.

    QUARTERBACKS. Howard Schnellenberger.

    DEFENSIVE SECONDARY. Bill Oliver.

    RECEIVERS. Jerry Sullivan.

    DEFENSIVE ENDS. Foge Fazio.

LINEBACKERS. Fred Goldsmith.

OFFENSIVE LINE. Marvin Johnson.

STRENGTH TRAINING. Bob Fix.

MENTAL GAME. Bob Goshen.

THE DEFENSIVE LINE. Defensive play for nose tackles, tackles and ends by NFL veteran Mike McCoy. (1987) 62 min., $29.95.

FOOTBALL WITH TOM LANDRY: *Quarterbacking to Win*. With Jim Zorn. 59 min.

HOW TO PLAY WINNING FOOTBALL. Instructional featuring Tom Landry, Jay Hilgenberg, Sonny Jurgensen, and Jim McMahon. $14.95.

KICKING GAME. Former Denver placekicker Rich Karlis on kicking and conditioning. $29.98.

LEARNING FOOTBALL THE NFL WAY: DEFENSE. Hosted by Phil Simms. 60 min., $19.98.

LEARNING FOOTBALL THE NFL WAY: OFFENSE. Hosted by Phil Simms. 60 min., $19.98.

KEN O'BRIEN'S QUARTERBACK CLINIC. 28 min., $14.95.

THE OFFENSIVE LINE. Center Pete Brock explains the basics of interior line play. (1987) 59 min., $29.98.

OFFICIAL POP WARNER VIDEO HANDBOOK. A basic instructional guide. (1988) 52 min., $19.95.

PLAYING TO WIN FOR BACKS AND RECEIVERS. Two-time Heisman Award winner Archie Griffin demonstrates proper techniques for running backs; all-time leading receiver Steve Largent shows how to get open and catch the ball. (MR, 1985) 59 min., $19.95.

PROFESSIONAL SPORTS TRAINING FOR KIDS: *Fooball with Dan Fouts*. Quarterback clinic with Fouts, plus conditioning workouts. 40 min., $19.95.

QUARTERBACK FUNDAMENTALS AND TECHNIQUES. Coach

Joe Kriviak with three of his star pupils, former U. of Maryland's Boomer Esiason, Stan Gelbaugh, and Frank Reich. (WKV) $14.95.

SPORTS CLINIC: FOOTBALL. Basics of line play, receiving and quarterbacking with George Allen, et al. (SCU) 80 min., $29.95.

SPORTS TRAINING CAMP: FOOTBALL. Brian Sipe hosts this basic instructional. 60 min., $19.95.

TEACHING KIDS FOOTBALL WITH BO SCHEMBECHLER. Michigan's coach on teaching beginning football. (ESP, 1986) 75 Min., $29.95.

VINCE LOMBARDI: *How to Play Winning Football.* Coach Lombardi and his Packer assistants cover the fundamentals of the game. 45 min., $19.99.

WINNING FOOTBALL WITH VINCE LOMBARDI. Detailed examination of each aspect of football. Six video set, each 30 min. Total 180 min., $99.95.

THE WINNING LINEBACKER. Former Dallas Cowboys star Bob Breunig shows how to read and react; former All-Pro cornerback Mike Haynes teaches the basics of defensive backfield play. (1986) 60 min., $19.95.

Bloopers and Humor

Though often imitated, NFL Films blooper videos are still the best football humor overall. The suggested price is $19.98, although some videos are twice as long as others. There are a few laughs among the imitators, but they don't have NFL Films' extensive film library to draw upon, or, in most cases, their sense of humor.

BEST OF THE FOOTBALL FOLLIES. The best (or worst) foulups, fumbles, and bloopers culled from 20 years of the successful videos by NFL Films. (NFL, 1985) 44 min., $14.98.

No less a personage than Wayne Gretzky (a.k.a. The Great One) was quoted last year: "I think it's time to take a hard look at what we are doing. People in the U.S. believe hockey is a violent sport because we allow fighting.

"I see guys like Brett Hull and Mario Lemieux and Steve Yzerman and others and wonder why the league isn't publicizing them more. I see talented Europeans entering this league, yet in order to keep them here and attract more talent from Europe, some rules are going to have to be changed.

"I'm not saying I'm right, that fighting should be eliminated, but that is the prevailing viewpoint of most Americans. We need to grow and we need the U.S. TV market. But we will not get on TV as long as we allow fighting."

Since the brawlers seldom are punished with more than a few minutes in the penalty box, I have to assume that the league tacitly condones such behavior as a way of putting fannies in the seats. But maybe it's time for the league to stop being pantywaist-wise and pounding-foolish. Take a chance on losing a few fisticuff fans and try for the larger and more lucrative TV market.

Interestingly enough, none of the videos I watched featured a single fist flung in anger.

Another criticism often raised is that just about every team ends up eligible for the playoffs—the "second season," as the NHL calls it. The NHL's playoffs are one of the most exciting spectacles in sports. Unfortunately, the inclusion of all but six teams tends to make the "first" season pretty small potatoes, and it's a little hard to get worked up over any one game.

Perhaps the biggest drawback hockey has in competing with baseball, basketball, and football is that it doesn't lend itself so well to television. Baseball is hardly the ideal TV game when something is happening, but there are so many pauses built into it that any crucial play can be shown again and again from a variety of camera angles. Football, with its start-stop pace, is perfect. But everything on the ice happens so blasted fast that there's little time for replays except after scoring plays. Admittedly, that can be a mixed blessing, especially because it also tends to reduce the time the announcer has to pontificate. Basketball has some of the same continuous-action problems as hockey, but it's played in a smaller space with a big, round ball that is visible even in the longest TV shots.

A large TV sin for hockey is that the puck is just plain hard to see. I follow it fine over the blue line, but then it gets down around the net and disappears among a blurry flurry of burly bodies until there's either another faceoff or the announcer yells, "He shoots and SCO-O-O-O-R-R-R-E-S!" How did he *see* that?

Happily, such invisible plays aren't found on videos where replays and slow motion make it possible to finally see what you missed on the night of the game.

In number available, if not in quality, hockey videos lag behind the other big team sports. Nevertheless, all of the categories can be found if you look hard enough: team highlights, season highlights, histories of some teams, individual stars, bloopers, and instructionals.

Pro Hockey Funnies will never win an award for truth in advertising. It's just not very funny. A "Hockey Opera" is a weak ripoff of NFL Films' "Hi, Mom. We're number one" showstopper. In trying to make its version funnier, the hockey video substitutes a flat soprano for the fine tenor used in the NFL classic. Instead of adding to the laughs her voice merely underlines how desperate the makers are to squeeze out a chuckle. Only slightly more amusing is a segment on practical jokes which in a hockey dressing room seem to consist mainly of surprising teammates with shaving cream.

Despite its lack of laughs, *Pro Hockey Funnies* has some worthwhile moments. Scattered through its 35 minutes are features on the Top 10 Goals, Top 10 Saves, and Top 10 Weird Goals. They aren't David Letterman-type Top 10's—no laughs even in the so-called "weird" goals—but they do make terrific highlights. There's one where the guy is on his knees with his back to the net and—well, you'll just have to see it to believe it.

The highlights in *Dynamite on Ice*, a legitimate highlight video, aren't any better than those Top 10's in *Funnies* but there are a few more of them. It includes segments on the 1989 All-Star Weekend, the game itself with Mario Lemieux's first period hat trick, other explosive endings, breakaways, and the Stanley Cup playoffs.

There's enough good stuff that we can forgive such grandiose narration as "sling-shotted an unstoppable projectile." That's probably why they put a mute button on your remote control.

The most interesting segment deals with the night Wayne Gretzky broke Gordie Howe's all-time NHL scoring record. To remind you, Gretzky went into the game needing one goal to tie Howe. He got it almost immediately, but as the clock was running out, his team trailed. Then, like Frank Merriwell, he slapped in the goal that tied the game and broke Howe's record. And THEN, he added to his record with the overtime goal that won the game. No wonder they call him "The Great One."

Howe, incidentally, hosts one of the most ambitious instructionals—a seven-part series that covers everything from conditioning to power skating. Most run about an hour and all are reasonably priced.

Hockey Videos

Highlights and History

The glory of highlights here, perhaps more than in any other sport, is slow motion. Plays that happen in a blink on ice can be studied and marveled at.

BLADES OF SUMMER. Highlights of 1987 Canada Cup. (1987) 60 min., $32.95.

C OF CHAMPIONS. 1989 playoff and Stanley Cup showdown won by Calgary Flames. (1989) 60 min., $19.95.

CANADA/RUSSIA GAMES 1972. 60 min., $32.95.

DON CHERRY'S ROCK 'EM SOCK 'EM HOCKEY. Collection of hits, saves, and goofs. (JCI, 1989) 33 min., $9.95.

DEVASTATING HITS IN HOCKEY. The super hits of five decades; hosted by Harry Neale. (QV, 1990) 45 min., $9.99.

DYNAMITE ON ICE. Face-to-face highlights from the 1990 NHL season. (PHO, 1990) 35 min., $9.95.

FANTASTIC HOCKEY FIGHTS. Sometimes sold as a two-for-the-price-of-one double feature with "Hilarious Hockey Highlights." 30 min., $9.95.

GREAT PLAYS FROM GREAT GAMES. Shocking comebacks, breathtaking finishes, and terrific shots. (JCI, 1989) 45 min., $9.95.

HOCKEY: A BRUTAL GAME. Rugged hits. 40 min., $9.99.

HOCKEY'S GREATEST HITS. Energetic collection of collisions. (SIM, 1989) 35 min., $9.95.

HOCKEY'S HARDEST HITTERS. Collection of hits by hockey heroes, including some cheap shots. (JCI, 1989) 30 min., $9.95.

HOCKEY SLAPSHOTS & SNAPSHOTS. Sold with 50 hockey cards. (JCI) $12.95.

LES CANADIENS. 75 year history of Les Habitants, the sport's "winningest team." 50 min., $39.95.

LIGHTNING ON ICE: *The History of Hockey*. Hosted by Alan Thicke. (1990) 50 min., $19.98.

LOS ANGELES KINGS 1988-89. Gretzky's first year with the Kings. (1989), $9.95.

LOS ANGELES KINGS 1989-90. (1990), $19.95.

(New York) RANGERS, STORY OF THE 88-89: YEAR OF THE ROOKIES. (1989), $19.95.

NEW YORK RANGERS HIGHLIGHTS 1989-90. (1990), $19.95.

ROUGH AND TOUGH HOCKEY: DON CHERRY VOL. 2. The sequel to Cherry's Rock 'Em Sock 'Em. (QV, 1991) 35 min., $9.99.

STANLEY CUP PLAYOFFS. Penguins defeat North Stars in Championship series. (1991) 60 min., $19.95.

A TIME TO REMEMBER. 1988 Stanley Cup highlight film. (1988), $19.95.

TRADITION ON ICE: *62 Year History of the New York Rangers*. (PAR, 1988) 60 min., $19.95.

Individual Stars

As you would expect, Gretzky and Lemieux are the hot tickets, but it's fun to watch Orr, Hull, Esposito, and some of the other stars of a few years back.

BOBBY CLARKE. Greatest Sports Legends. (RVI) 30 min., $9.95.

PHIL ESPOSITO. Greatest Sports Legends. (RVI) 30 min., $9.95.

SPORTS CHAMPIONS: WAYNE GRETZKY. 50 min., $9.95.

WAYNE GRETZKY: *Above and Beyond.* Hockey's greatest star. (IVE, 58 min., $19.95.

HOCKEY SUPER STARS. Orr, Hull, Howe, Esposito. 90 min., $19.95.

GORDIE HOWE. Greatest Sports Legends. (RVI) 30 min., $9.95.

BOBBY HULL. Greatest Sports Legends. (RVI) 30 min., $9.95.

MARIO THE MAGNIFICENT. The career of Mario Lemieux on and off the ice. (QV, 1991) 45 min., $19.95.

MARIO LEMIEUX: SUCCESS STORY. The life and mostly good times of Super Mario. (1992) 50 min., $9.95.

BOBBY ORR. Greatest Sports Legends. (RVI) 30 min., $9.95.

Instructionals

Hockey instructionals will become more common as the sport keeps growing on a junior high and high school level. At this point personable Gordie Howe's series is the most accessible.

HOCKEY FOR KIDS AND COACHES. Fast-paced instructional with Doug Wilson, Bob Gainey, Rod Langway, and more. 60 min., $19.95.

HOCKEY: HERE'S HOWE, CONDITIONING AND COACHING. Instruction from Hall of Famer Gordie Howe. (KR, 1989) 55 min. $14.95.

HOCKEY: HERE'S HOWE, DEFENSE. (KR, 1989) 55 min., $14.95.

HOCKEY: HERE'S HOWE, FORWARDS. (KR, 1989) 55 min., $14.95.

HOCKEY: HERE'S HOWE, GOALTENDING. (KR, 1989) 60 min., $14.95.

HOCKEY: HERE'S HOWE, POWER SKATING. (KR, 1989) 35 min., $14.95.

HOCKEY: HERE'S HOWE, SHOOTING. (KR, 1989) 50 min., $14.95.

HOCKEY: HERE'S HOWE, STICK HANDLING AND PASSING. (KR, 1989) 35 min., $14.95.

NEAL BROTEN'S GOLD MEDAL HOCKEY. Fundamentals of hockey. 60 min., $39.95.

WAYNE GRETZKY: Hockey My Way. Basic techniques and the finer points of the game. $39.95.

Bloopers and Humor

Hockey blooper videos aren't generally as funny as some other sports yet, but the elements are there, and they should get better.

HILARIOUS HOCKEY HIGHLIGHTS. Sometimes sold as a two-for-the-price-of-one double feature with "Fantastic Hockey Fights." 30 min., $9.95.

HOCKEY: THE LIGHTER SIDE. Bloopers, spills, bonecrushing checks, etc. (SIM, 1988) 30 min., $9.99.

SUPER DOOPER HOCKEY BLOOPERS. Flubs, falls, and foulups. 35 min., $9.95.

PRO HOCKEY FUNNIES. More highlight than ha-ha. (PHO, 1990) 35 min., $9.95.

SOCCER

It seems like only yesterday—well, the day before yesterday—that soccer aficionados were predicting the kicking game was about to take the United States by storm and make everybody forget all about baseball, basketball, and football. Soccer is the most popular game in the world, went the reasoning, and Americans are no different from people in other countries. Once they get a good dose of soccer, they'll be terminally infected. As it turned out, America's bout with soccer has been more like a light cold.

The North American Soccer League was founded in 1968 and gave up the ghost in 1985. Even bringing in soccer immortal Pélé couldn't save it. The Major Indoor Soccer League began in 1979 and is still around, but if one of its eight teams isn't in your city you may not know the league exists. And no matter where the teams are, the majority of America's sports fans couldn't care less.

Nevertheless, all is not dark on the hockey horizon. For one thing, the World Cup, soccer's championship, will be held in the U.S. in 1994. On a planetary scale, the World Cup is bigger than the Olympics. Although many critics from other nations have suggested that having the World Cup decided in soccer-blasé America is like holding the Super Bowl in Estonia, U.S. soccer fans (and there are some) hope that the resulting publicity will help spring their favorite sport into American consciousness.

Perhaps a more long-lasting influence on soccer's U.S. future will be the growth of the game in American schools. All the way from junior high through college, it's becoming more popular to play and to watch. While most of this is due to the game's inherent charms, some of the growth must be ascribed to the fact that soccer is both cheaper and safer to play than football. As costs for equipment and insurance rise in football, more and more schools are finding soccer an attractive alternative.

Those who still predict a bright future for soccer in this country maintain that eventually all those kids who grow up playing soccer will form a vast viewing audience as they approach middle age. "Vast" may be overly optimistic, but "bigger" seems very possible.

On the other hand, it seems unlikely that the majority of Sunday football fans who fill the NFL's stadiums and keep its Neilsen ratings high ever played much football themselves. If soccer is one day to rival football as a spectator sport, it will probably have to do so as a spectacle. Maybe, for North American fans, they should add a note of peril... pointy-toed shoes or something. There's a lot of peril involved with soccer in other countries, but it's all in the stands.

Because soccer interest in North America is to be found primarily among young players, nearly every soccer video available here is an instructional. If you're just a raw beginner, something like *Teaching Kids Soccer with Bob Gansler* may be just what you're looking for. Gansler, coach of the U.S. Soccer Federation's National Team, teaches basics plus drills to improve skills. It pays attention to injury prevention, too, which will gladden any mother's heart.

Coaching Youth Soccer is the Official U.S. Youth Soccer Association's Coaching Guide for ages 6 to 9.

Graduated Soccer Method Series: Fundamentals and Techniques is a three-part series of self-training tapes for kids ages 7 to 14. You start with *Building a Relationship with Your Soccer Ball*, a 30-minute love story that should improve kids' confidence with the one essential piece of equipment. Next comes *Developing Fast Feet / Shielding the Ball,* wherein speed and coordination are developed. And finally, *Taking on Your Opponent*, which discusses strategy.

A more detailed set of videos is *Videocoach Vogelsinger's Soccer Series*. Five videos, each about 100 minutes, guide a young player through soccerobics, kicking, dribbling and feinting, ball control, and super skills and heading.

If you want to see and hear how soccer's greatest star did it, there's *Pélé Soccer Training Program: The Master and His Method*. Although Pélé goes through nearly every aspect of soccer play, you still might feel that there was more to his brilliance than can be expressed in 60 minutes.

I found only one highlight video available at this time, *Goals Galore*. (Wasn't she a character in a James Bond movie?) It contains over 100 often-spectacular goals.

I looked in vain for soccer blooper tapes. No doubt such are available in Europe, but that's outside the range of this book. However, if you happen to come up with one for your kid, you won't have to translate "Es ist der soccer ball." Every place else except the North American continent, they call the game football. Of course, in German, that's *fussball*.

The reason we use "soccer" in America is that about 100 years ago the newspapers usually referred to the game as "Association Football" to distinguish it from our "American" football. After a while, they started just abbreviating it as "Assoc" which somehow got transmogrified into "soccer." At least, that's the explanation that's always given.

Soccer Videos

Highlights

Although soccer highlight videos are still rare in this country, there is a strong likelihood that quite a few will be added in the '90s as more and more former high school players grow to maturity.

GOALS GALORE: *Over 100 of Soccer's Greatest Goals.* Scored to rock music. 20 min., $9.98.

Individual Stars

Even today, after two decades, the only soccer star most U.S. non-soccer fans can name is Pélé.

PELE. Greatest Sports Legends. (RVI) 30 min. $9.95.

Instructionals

That nearly all available soccer videos are instructionals is indicative of the progress the game has made among the young and, perhaps, that many coaches charged with leading scholastic teams are not former players and need instruction themselves.

COACHING YOUTH SOCCER: *Official U.S. Youth Soccer Assoc. Coaching Guide for Ages 6-9.* 50 min., $24.95.

DO IT BETTER: SOCCER. Asics Sports Video Collection; hosted by Coach John Boyle. 30 min., $24.95.

GRADUATED SOCCER METHOD SERIES: *Fundamentals and Techniques*; self-training tapes for boys and girls, ages 7-14 (WKV)

VOL. I: BUILDING A RELATIONSHIP WITH YOUR SOCCER BALL. Developing confidence and basic skills. (1986) 50 min., $14.95.

VOL. II: DEVELOPING FAST FEET / SHIELDING THE BALL. Improving speed, coordination, and ball control. (1986) 30 min., $14.95.

VOL. III: TAKING ON YOUR OPPONENT. Various techniques that can be used to outwit an opponent. (1986) 30 min., $14.95.

HEAD TO TOE: *Soccer for Little Leaguers.* Basic training for junior players, starring Wayne Jentas. (1988) 35 min., $19.95.

HOT TIPS SOCCER SERIES: *Bobby Charlton's Soccer, Level 1.* A quick tour through discipline, dress, and skills by England's famous coach. (1989) 30 min., $9.95.

HOT TIPS SOCCER SERIES: *Bobby Charlton's Soccer, Level 2.* Improving skills and applying them to game situations. (1989) 30 min., $9.95.

PELE: *The Master and His Method.* Pélé's complete program on ball control, passing, dribbling, conditioning, etc. (1987) 60 min., $19.98.

SOCCER FUNDAMENTALS. Individual skills taught by Shep Messing. (1988) 30 min., $29.95.

SOCCER SERIES: *Basic Individual Skills, Offensive and Defensive Maneuvering, Goal Keeping.* 61 min., $29.95.

SOCCER TACTICS AND SKILLS. Taught by Coach Charles Hughes along with English stars Peter Shilton and Kevin Keegan.

THE ATTACK. Explains how to increase your team's "momentum of attack" with sound tactics. (1988) 60 min., $29.95.

CREATING SPACE. Basics of ball control and the art of creating and exploiting space. (1988) 60 min., $29.95.

DEFENDING. Mental concentration, patience and discipline on defense. (1988) 60 min., $29.95.

GOALKEEPING. Techniques in keeping opponents from scoring. (1988) 60 min., $29.95.

PASSING AND SUPPORT. Improving ground passes and many others through improved techniques. (1988) 60 min., $29.95.

SET PLAYS. Attacking and defending set plays. (1988) 60 min., $29.95.

SHOOTING. How to improve scoring through a variety of demonstrations. (1988) 60 min., $29.95.

SOCCER WITH THE SUPERSTARS. Bernie James and Peter Word teach individual skills. 30 min., $9.95.

SPORTS CLINIC: SOCCER. With Coach Hubert Vogelsinger. (SCU, 1987) 80 min., $19.99.

SPORTS TRAINING CAMP: SOCCER. With Chuck Clegg. 60 min., $19.95.

STRIKER TACTICS: *Skills to Help You Score, Part I.* England's Coach Bobby Charlton in a two-part series. (1988) 35 min., $19.95.

STRIKER TACTICS: *Skills to Help You Score, Part II.* England's Coach Bobby Charlton in a two-part series. (1988) 30 min., $19.95.

TEACHING KIDS SOCCER WITH BOB GANSLER. Teaching children the fundamentals of soccer. (ESP, 1987) 75 min., $29.95.

VIDEOCOACH VOGELSINGER'S SOCCER SERIES

VOL. I: SOCCEROBICS. 100 min., $29.95.

VOL. II: KICKING. 100 min., $29.95.

VOL. III: DRIBBLING AND FEINTING. 100 min., $29.95.

VOL. IV: BALL CONTROL. 100 min., $29.95.

VOL. V: SUPER SKILLS AND HEADING. 100 min., $29.95.

WINNING SOCCER: *Basics of the Game.* Dribbling, passing, shooting, receiving, and goalkeeping; featuring Shep Messing. 47 min., $14.95.

BOXING

In **October 1991**, the prestigious *American Heritage* magazine featured a boxing article and cover. The editors received the following:

> I am angry.
> I was shocked and ashamed…. The same *civilized* society that forbids bullfights, dogfights, cockfights, and bearbaiting encourages and highly remunerates human beings to maim, cripple, and occasionally kill one another. The promoters and participants should be charged with felonious assault and battery.

If you're old enough, you may remember Red Skelton's hilarious routine about a punch-drunk fighter. There was a time when just about everybody laughed at "Boy! A flock of 'em flew over that time!" In our more enlightened age, we don't find the brain-damaged amusing. We no longer laugh at racial stereotypes or drunks either. Even mother-in-law jokes are an endangered species. Maybe we've gotten smarter. Maybe we've just seen too many tragedies.

I wasn't able to find a "Boxing's Biggest Bloopers" video. Probably such a thing would be considered in bad taste by those who lump the sweet science with bearbaiting and cockfighting. All the boxing videos I found were either general highlights of various bouts or highlights from individual careers. If there was one single strain running through them, it was the knockout.

Among the general videos available are *Boxing's Greatest Knockouts, Vol. I, Mike Tyson and History's Greatest Knockouts*, and *The

Heavyweights: The Big Punchers. Videos dealing with individual careers nearly all celebrate men with iron in their hands: Jack Dempsey, Joe Louis, Rocky Marciano, Muhammad Ali, and Mike Tyson.

And that tells us something about most people who watch boxing—they want to see one guy or the other knocked silly. Certainly there are boxing fans who find their greatest joy in seeing a clever boxer outpoint his opponent with snappy jabs and dazzling footwork. There just aren't enough of them to make up a profitable video market. What sells videos is the same thing that puts fannies at ringside at five hundred bucks a seat—the prospect of ka-boom!

Such bloodlust in boxing fans is sometimes equated with the joy that football fans find in a big hit or the fascination that auto racing fans feel for an accident, but that's an unfair simplification for all three groups. Football fans will always cheer louder for a long scoring play than for a hard tackle, and they are horrified when a rugged hit results in a serious injury. Likewise, no true racing fan wants to see a crash. To say we watch the crippling hit or flaming wreck on the eleven o'clock news is only to say that we all are morbidly fascinated with brutal action; it doesn't mean we'd vote for it.

The appeal of the knockout in boxing is not the damage wrought; it's the finality of it. It ends the contest immediately. In what other popular sport can the drama be suddenly resolved at any moment along its continuum? A first- inning home run, a second-quarter basket, a third period goal... what happens? The game goes on. Only in overtime among the major team sports can you find the sudden, decisive finality of a knockout. One moment the contestants are toe-to-toe and the next moment (or more precisely, ten seconds later) the competition has been resolved.

Boxing has its critics, and they make good points. It is brutal, and physical damage is done to the participants. Sometimes the damage is permanent. Physical impairment is a possibility in every sport, but even an NFL veteran's plastic knees don't compare in severity with the damaged brains of some ex-boxers.

It's also true that much of the promotional side of boxing is hip-deep in sleaze. Many believe that the sins of promoters are primarily responsible for most of the physical ills of the boxers.

Boxing videos ignore the dark side of boxing, of course. Yet, oddly, most of the successful theatrical movies with boxing backgrounds *stress* the ugly elements. One wonders if the same person who watches *Boxing's Greatest Knockouts, Vol. I* on his VCR would later tune in *The Harder They Fall* on the late show.

Despite the validity of criticisms, boxing remains extremely popular in this country. Million dollar gates for big fights are common. A televised championship bout invariably draws high ratings. Does it then follow that Americans are a brutal people, no better than the Romans cheering on the lions? Hardly.

Before we stir too much *mea* into our *culpa*, we should note that most foreign boxing champions are bigger heroes in their own country than American champions have been in this one. And the twentieth century's most popular athlete worldwide has been a boxer—Muhammad Ali.

On the other hand, while boxing maintains its popularity as a spectator sport, I found only a single instructional video, *Emile Griffith's Learn to Box*. Does this mean that few papas want to raise their boys to be punchers? Or is it simply that those in the lower economic levels—traditionally where boxing has found its most hopeful aspirants—do not constitute a major video market?

Boxing Videos

Highlights, General

Among the earliest sporting sporting events filmed were big fights. As a consequence, nearly every important match of the twentieth century may eventually end up on video if there's a market. Right now, bouts involving Ali, Tyson, and Leonard are the best represented.

BAER VS. LOUIS 1935 / LOUIS VS. SCHMELING 1936. Two of the Brown Bomber's key fights on his way to the title. 54 min., $24.95.

BOXING'S GREATEST CHAMPIONS. Boxing's Best Series; Marciano, Ali, Louis, Duran and others. (HBO, 1990) 60 min., $19.99.

BOXING'S GREATEST KNOCKOUTS, VOL. I. The precise moments when victory was achieved in many of the biggest fights ever. (NAC, 1989) 40 min., $9.95.

BOXING'S GREATEST KNOCKOUTS AND HIGHLIGHTS. A tribute to punching superstars in some of the greatest fights of the century. 30 min., $9.95.

BOXING'S GREATEST UPSETS. When the underdog shocked the world; Norton over Ali, Ali over Spinks, Leonard over Hagler, Barkley over Hearns. (1991) 64 min., $19.98.

CHAMPIONS FOREVER. Ali, Foreman, Frazier, Holmes, Norton tell their stories. (J2, 1989) 87 min., $19.95.

THE FABULOUS FOUR. The four-way rivalry of Leonard, Duran, Hearns, and Hagler. (CMV, 1991) 83 min., $19.98.

GREATEST ROUNDS EVER, PART 1. Dempsey-Willard, Louis-Schmeling, Marciano-Walcott, others. 68 min., $9.95.

GRUDGE FIGHTS. Boxing's Best Series; when more was on the line than the purse, from Louis-Schmeling to Ali-Frazier. (HBO, 1990) 58 min., $19.99.

THE HEAVYWEIGHTS: *The Big Punchers*. Boxing's Best Series; knock-out artists, including Dempsey, Louis, Marciano, Frazier. (HBO, 1990) 59 min., $19.99.

THE HEAVYWEIGHTS: *The Stylists*. Boxing's Best Series; eight of the all-time best and their unique styles. (HBO, 1990) 60 min., $19.99.

LEGENDARY CHAMPIONS. A nostalgic look at boxing from the 1890s to 1929. (HBO, 1990) 85 min., $19.99.

LEGENDS OF THE RING. History of boxing from ancient times through the twentieth century. 65 min., $29.95.

MIDDLEWEIGHTS. Boxing's Best Series; from Graziano and Robinson to Monzon and Hagler. (HBO, 1989) 60 min., $19.99.

TEN AND COUNTING: *Best of ESPN Boxing*. Features the TV debuts of Tyson, Douglas; Foreman's 1988 comeback; Czyz, others. (ESP) 45 min., $9.95.

30 GREAT ONE-PUNCH KNOCKOUTS. When a single blow ended it all. (1991) 30 min., $19.95.

30 MORE ONE-PUNCH KNOCKOUTS. (1991) 64 min., $19.95.

RINGSIDE WITH MIKE TYSON. Boxing's Best Series; Tyson recalls boxing's finest moments: upsets, controversies, and comebacks. (HBO, 1990) 60 min., $19.99.

TYSON'S GREATEST HITS. "Iron Mike's" knockout route to the

championship. (HBO, 1988) 60 min., $19.99.

MIKE TYSON AND THE HEAVYWEIGHTS. Boxing's Best Series; Tyson's history of the heavyweight division, including rare footage. (HBO, 1990) 52 min., $19.99.

MIKE TYSON AND HISTORY'S GREATEST KNOCKOUTS. Boxing's Best Series; boxing's most stunning endings, including Louis-Schmeling, Ali-Liston and others. (HBO, 1990) 58 min., $19.99.

Highlights, Individual

ALI-FOREMAN: THE RUMBLE IN THE JUNGLE. One of "The Greatest's" greatest triumphs. 80 min., $9.95.

ALI-FRAZIER: THE THRILLA IN MANILLA. Ali called it the closest he's come to death. 60 min., $9.95.

ALI-NORTON TRILOGY. All three meetings between Muhammad Ali and Ken Norton. (NAC, 1989) 74 min., $9.95.

MUHAMMAD ALI. Boxing's Best Series; an hour-long tribute to the most charismatic boxer in history. (HBO, 1990) 58 min., $12.98.

MUHAMMAD ALI: BOXING'S BEST. Highlights of his storied career. 30 min., $25.00.

MUHAMMAD ALI VS. ZORA FOLLEY: MARCH 22, 1967. 15 round title fight from Madison Square Garden, with pre- and post-fight interviews. 68 min., $24.95.

CLAY-LISTON: CLAY SHOCKS THE WORLD. The brash youngster becomes the heavyweight champ. 45 min., $9.95.

JACK DEMPSEY. Boxing's Best Series; action from the career of one of history's most ferocious punchers. (HBO) 45 min., $19.99.

POWER PROFILES: JACK DEMPSEY. The Manassa Mauler's career. 24 min., $19.95.

DOUGLAS-TYSON: *The Upset of the Century.* A 42-1 underdog stops the unstoppable. 93 min., $19.99.

JOE FRAZIER. Greatest Sports Legends. (RVI) 30 min., $9.95.

JACK JOHNSON. Boxing's Best Series; archival footage of the controversial champion both in and out of the ring. (HBO, 1990) 55 min., $19.99.

SPORTS CHAMPIONS: SUGAR RAY LEONARD. From his Olympic gold medal to his first retirement. 50 min., $21.50.

THE LEONARD VS. HEARNS SAGA. Includes both their 1981 meeting and the 1989 rematch. (CBS, 1989) 118 min., $19.98.

LEONARD-DURAN III: *"Uno Mas."* The third installment of one of the premier series of matches in the 1980s. (NAC, 1989) 70 min., $14.98.

JOE LOUIS. Boxing's Best Series; the Brown Bomber became one of the greatest champions of all time. (HBO, 1989) 60 min., $19.99.

JOE LOUIS: *Fort All Time.* His life and career. 87 min., $29.95.

POWER PROFILES: JOE LOUIS. Life and career. 24 min., $19.95.

ROCKY MARCIANO. Boxing's Best Series; the Brockton Bomber was one of the sport's greatest knockout kings. (HBO, 1990) 60 min., $19.99.

SUGAR RAY ROBINSON. Boxing's Best Series; "pound for pound" perhaps the greatest fighter who ever lived. (HBO, 1990) 60 min., $19.98.

Instructional

EMILE GRIFFITH'S LEARN TO BOX. 65 min., $39.99.

TENNIS

Tennis has its roots in royalty, a version having been played in royal courts of fifteenth-century France. Yes, that's apparently why one plays the game today on a "court" rather than a "field." It's also probably why the ball is "served" instead of "hit-like-hell." A modern version of the game was invented by an Englishman named Wingfield in 1873 and was being played in this country within a year. Although America lacked true bluebloods, our version of royalty—the rich—were the first to embrace the new game. The middle and lower classes lacked the space and leisure time to pursue it with any dedication.

Professionalism—the idea of earning one's daily bread by playing a game—came to tennis rather late: 1926, to be exact. By that time, baseball players had been paid for more than sixty years. Generally, until the last thirty years or so, tennis was still considered a polite activity for, if not only the rich, at least the well-to-do. And although some racketeers achieved national stardom—Bill Tilden, Jack Kramer, Pancho Gonzales, Alice Marble, and Maureen Connolly, to name a few—the game was played and viewed for the most part by a polite minority of the sporting crowd.

But, like that skinny Virginian, tennis has come a long way, baby. Today it is big business and a good tennis pro can earn more in a few tournaments than Tilden earned in his career. Perhaps more importantly, it's become a leisure pursuit for millions who are just happy to get the ball over the net one out of three times. And, through television, millions more who would never think of embarrassing themselves on an actual court can critique the best efforts of Connors, Lendl, Agassi, Navratilova, and a host of others.

Tennis videos are available in three flavors: general highlights, instructionals, and a very few that focus on individual stars. The instructionals are by far the most common.

Tennis Videos

Highlights

Most highlight videos naturally focus on the past decade's superstars, but don't ignore such tapes as "Smashing Ladies" which will give you the chance to see how the game has changed.

BEST OF U.S. OPEN TENNIS. Highlights classic matches of the 1980s. 45 min., $19.98.

THE OFFICIAL 1989 U.S. OPEN TENNIS VIDEO. The biggest stars of the game, plus a special tribute to Chris Evert. (CBS, 1989) 60 min., $19.98.

THE 1988 U.S. OPEN VIDEO. Graf, Navratilova, and Lendl all bid for success, plus a special tribute to the tournament's twentieth anniversary. (CBS, 1988) 60 min., $19.98.

GOLDEN GREATS OF TENNIS. $14.95.

SMASH HIT. Highlights and candid interviews with Connors, Lendl, Evert and others. 70 min., $19.95.

SMASHING LADIES: *The Legends of Women's Tennis*. Women tennis greats from the 1920s through the '60s. 60 min., $19.95.

WIMBLEDON: *The One to Win*. Traces the history of the most prestigious tournament in tennis. (HPG, 1989) 59 min., $19.95.

Instructionals

There are more than enough good instructionals available; a good place to start is "Body Prep." One good thing about taking lessons from McEnroe, Connors, or Lendl on tape: it won't elicit a temper tantrum from the teacher.

ATTACK. Andre Agassi and coach Nick Bolletieri use slow motion and graphics to show how to improve your stroke. (NAC, 1990) 60 min., $29.95.

BODY PREP, TENNIS! Olympic silver medalist Tim Mayotte and others in a two-part video showing basic exercises and follow-along workouts to get ready for tennis. (WMH, 1990) 106 min., $29.95.

VIC BRADEN TENNIS

>VOLUME 1. Braden covers the four fundamentals: forehand, backhand, serve and volley. (PAR, 1983) 108 min., $14.95.
>
>VOLUME 2. The approach shot, spin and service return, the overhead, lob, drop shots, and conditioning. (PAR, 1983) 105 min., $14.95.
>
>VOLUME 3. Singles strategey, doubles, and psychology of the game. (PAR, 1983) 104 min., $14.95.

JIMMY CONNORS' TENNIS: *Winning Fundamentals*. Comprehensive instruction from today's most popular player; Connors demonstrates the basics of a strong ground game. (VES, 1988) 60 min., $39.98.

JIMMY CONNORS: *Match Strategy*. Connors on techniques and strategy to sharpen skills. (VES, 1988) 60 min., $39.98.

MAXIMIZING YOUR TENNIS POTENTIAL WITH VIC BRADEN. Improving your game with Braden's approach. 42 min., $29.95.

JOHN McENROE AND IVAN LENDL: *The Winning Edge*. Step-by-step instruction on basic tennis with the game's best. 45 min., $19.95.

TEACHING KIDS TENNIS. Especially made to teach the very

young the basics, featuring Nick Bollettieri. (ESP, 1988) 75 min., $29.95.

TENNIS FROM THE PROS. Jack Kramer, Arthur Ashe, Stan Smith teach techniques. $9.95.

TENNIS OUR WAY. Vic Braden, Arthur Ashe, and Stan Smith give pointers. (1987) 151 min., $39.95.

TENNIS TO WIN. With John Newcombe and Bjorn Borg

    VOL. I. Forehand, backhand, net play, and serving. 65 min., $16.95.

    VOL. II. Approach shots, lobs, drop shots, drop volleys, tactics, and strategy. 65 min., $16.95.

TENNIS WITH VAN DER MEER

    VOL. 1, ESSENTIAL STROKES, THE BASIC GAME. 60 min., $29.95.

    VOL. 2, THE ATTACKING GAME. 60 min., $29.95.

    VOL. 3, STRATEGY, TACTICS AND THE MENTAL GAME. 60 min., $29.95.

VIRGINIA WADE'S CLASS. The 1977 Wimbledon winner gives tips for beginners and veterans. (1988) 67 min., $14.95.

VISUAL TENNIS. Improving your game through visualization and mental imagery. 60 min., $29.95.

WARM UP TO ATTACK. Getting your body and mind ready to compete. 30 min., $9.95.

Individuals

This is definitely an understocked field. Where are Lendl, Becker, Ashe? Where are Graf, Natratilova, and King?

BJORN BORG. Greatest Sports Legends. (RVI) 30 min., $6.98.

JIMMY CONNORS. Greatest Sports Legends. (RVI) 30 min., $6.98.

GOLF

Frustration, thy name is golf.
It looks so easy! After all, the ball can't hit back. All you have to do is smack it with a stick a couple of times until it goes into a hole.

Of course, anyone who's ever tried it knows just how maddening the game can be. No doubt every golfer from the greatest pro to the dreariest duffer has uttered those immortal words: "I hate this game!" You can even find a video with that title—albeit one of the humorous ones.

Despite all the frustrations, golf is really a wonderful game. No matter what your age, build, or sex, you can spend an enjoyable afternoon walking over real green grass and hitting that ball. John D. Rockefeller was still playing when he was in his nineties. Some great golfers are small and wiry; some are as round as the ball. Many of the finest golfers in the land are women. Not even tennis can be played and enjoyed by such a wide range of people.

Few of us will ever compete in a major tournament against PGA pros, but we all can take a shot at beating our most important opponent—ourselves. Quite naturally, the most common and best-selling golf videos are instructionals. You can take lessons in your living room from Nicklaus, Palmer, Lopez, Trevino or just about any great golfer you can name.

Although actual figures are impossible to come by, the best-selling instructional is said to be the two-part *Jack Nicklaus' Golf My Way*. It is by far the most expensive, with each part running in excess of two hours and each costing about $85. However, if you're just starting out and aren't certain how far you want to go with golf, you should be aware that there are any number of excellent instructionals by lesser lights available for about $30 or even $20. Among the more popular tapes for beginners are *Bob Rosburg's Break 90 in 21 Days* for men and *Judy Rankin's Break 90 in 21 Days* for women. Many of us would be thrilled to break 90 on the front nine! Lee Trevino's numerous videos have the added benefit of his inimitable sense of humor.

Some of the best teachers are not major names on the tour. Dick Lawrence and Joyce Ann Jackson of the Ben Sutton Golf School are teaching pros who can be seen in *Golf Fundamentals* and *Women's Golf Fundamentals*, respectively.

When it comes to humor, golf ranks second in quantity only to baseball among sports. Indeed, there are no doubt more jokes about golf than about baseball. Visual humor, however—the kind you look for in videos—is not quite so prevalent. After all, how many slices and muffed shots can you take? Nevertheless, there are several chuckles available in various blooper-type videos. And, although Tim Conway's zany comedy is not to everyone's taste, he is to mine. I enjoyed both *Dorf on Golf* and *Dorf's Golf Bible*.

Some people—nongolfers for sure—say that the term "golf highlights" is an oxymoron. On the eleven o'clock news, you'll seldom see much more than the final putt for even the largest tournaments, and that's shown only because it caps the tournament. When a tournament comes on the TV in a bar, someone will always suggest switching channels. ("See what's on PBS fer Godsake, Harry!")

It's true that much that could be exciting to watch in person does not translate well to the TV screen. It's hard to get a sense of distance for a drive or of how well an iron shot is placed. The absolute worst are those seemingly endless pictures of a white ball moving against an empty blue sky. Putts are better, but they stand around so long before they try them!

Of course, what constitutes a highlight is more than a good shot; it's making that shot under pressure. Several historical highlight videos on such tournaments as the Masters and the U.S. Open make that point pretty well, and we prefer them generally to videos composed of spectacular, but not as pressure-packed, drives, chips, and putts.

Golf Videos

History and Highlights

Here they are! All those shots that made the galleries cheer and Pat Summerall whisper! Vids like "Fabulous Putting" are okay as highlights, but we prefer historical videos like "U.S. Open: Golf's Greatest Championship" where the context adds drama to even ordinary shots.

DALY'S ... THE LONG SHOT: *1991 PGA Championships*. He wasn't even in the original field but he won. (1992) 52 min., $29.95.

FABULOUS FINISHES OF THE PGA TOUR. From Palmer's charges of the 1960s up to today's duels. (1992) 43 min., $19.95.

FABULOUS PUTTING. A highlight video of putting only; the longest, hardest, etc. (PAR, 1988) 40 min., $19.95.

GOLDEN GREATS OF GOLF. Highlights of golf's all-time greatest players. Hagen, Jones, Sarazen, Nicklaus, Trevino. (GD) 60 min., $14.95.

GOLDEN GREATS OF GOLF. Greatest Sports Legends. Sarazen, Jones, Hogan, up to today. (RVI) 45 min., $22.95.

GOLF DIGEST VIDEO ALMANAC 1989. Highlights of the 1989 PGA Tour. (GD, 1989) 60 min., $14.98.

GOLF'S GREATEST MOMENTS. 100 years of American golf; includes rare archival film. (VES, 1989) 77 min., $29.98.

GOLF'S GREATS: VOL. I. Highlights from the careers of great players. (RVI) 30 min., $6.95.

GOLF'S ONE IN A MILLION SHOTS. Some of the greatest shots from some of the biggest tournaments. (WKV) 60 min., $29.95.

GREAT MOMENTS OF THE MASTERS. History of the Masters Tournament. (HPG, 1989) 54 min., $49.95.

HISTORY OF THE PGA TOUR. Historical footage and exclusive interviews. (1990) 76 min., $19.95.

HISTORY OF THE RYDER CUP. Seven decades of a true golf classic. 81 min., $24.95.

IMAGINE ALL EAGLES. Compendium of one of golf's greatest achievements. (PAR, 1990) 40 min., $19.95.

THE 1960 MASTERS TOURNAMENT. The first Masters to be captured in color had Palmer coming from behind, plus Hogan, Snead, Player, and Nicklaus. (RSF) 60 min., $29.95.

MASTERS TOURNAMENT 1986: *A Golden Moment in the History of Sports.* Nicklaus's magnificent comeback. (HPG, 1986) 60 min., $19.95.

MASTERS TOURNAMENT 1987. Three-way battle among Mize, Norman, and Ballesteros. (HPG, 1987) 60 min., $19.95.

MASTERS TOURNAMENT 1988. Curtis Strange's hole-in-one, Norman's late charge, etc. (HPG, 1988) 60 min., $19.95.

MASTERS TOURNAMENT 1989. From Trevino's 67 to Faldo's sudden-death win. (HPG, 1990) 52 min., $19.95.

MASTERS TOURNAMENT 1990. Faldo's playoff triumph. (HPG, 1991) 52 min., $19.95.

MASTERS TOURNAMENT 1991. Woosnam, Watson and Olazabal battle to the last stroke. (HPG, 1992) 52 min., $19.95.

THE MASTERS. History of the tournament through 1992 [see Great *Moments of the Masters* above]. (HPG, 1992) 60 min., $49.95.

101 GREAT PUTTS. Highlight video of the greatest putts ever. (1992) 49 min., $22.95.

RYDER CUP 1991. A dramatic charge to victory. (1992) 62 min., $29.95.

SUNTORY WORLD MATCH PLAY. World Match Play Championships at the famous Wentworth Golf Club in England since 1964. 55 min., $24.95.

TEN YEARS OF THE BRITISH OPEN: THE 1980S. The shots that won the tournament in the decade. (1990) 52 min., $24.95.

U.S. OPEN: *Golf's Greatest Championship*. The most spectacular moments in U.S. Open history. (VES, 1988) 60 min., $24.95.

U.S. OPEN 1987. Jack Nicklaus seeks his fifth U.S. Open but the showdown is Simpson-Watson. (1988) 60 min., $29.95.

U.S. OPEN 1988. Curtis Strange over Nick Faldo. (1989) 30 min., $19.95.

U.S. WOMEN'S OPEN 1988: A STAR IS BORN. Record-breaking performance of Liselotte Neumann. (1989) 55 min., $19.95.

Famous Courses

BAY COURSE AT KAPULUA. Golfing at an island paradise. 53 min., $17.95.

BULLYBUNION: THE OLD COURSE. High above the Atlantic on the rugged Kerry coast. 60 min., $29.95.

CLASSIC GOLF EXPERIENCES. An eight-tape extravaganza with each video taking you hole-by-hole through one of the world's greatest courses. (J2) $199.95.

DORAL'S BLUE MONSTER. Hole-by-hole through the home of the Doral Ryder. 44 min., $17.95.

GREAT GOLF COURSES OF SOUTHERN CALIFORNIA. TPC Stadium Course, Industry Hills, Ojai Valley Club, Torrey Pines, many more. 55 min., $24.95.

GREAT GOLF COURSES OF THE WORLD: SCOTLAND. St. Andrews, Muirfield, Carnoustie, many more. Narrated by Sean Connery. 77 min., $29.98.

GREATEST 18 HOLES OF CHAMPIONSHIP GOLF. Nicklaus travels the world to show, describe, and play the best. (CBS, 1986) 60 min., $19.98.

HARBOUR TOWN. Home of the Heritage Classic. 52 min., $17.95.

HISTORY AND TRADITIONS OF GOLF IN SCOTLAND. The world's most famous courses, including St. Andrews and the Links, are explored. 45 min., $9.95.

LA COSTA. Home of the Tournament of Champions. 40 min., $17.95.

MYRTLE BEACH GOLF: *Great Golf Courses You Can Play!* Covers six Myrtle Beach courses. 50 min., $24.95.

PGA TOUR: *18 Toughest Holes.* A hole-by-hole tour of the most notorious and toughest. (1992) 45 min., $19.95.

PGA WEST STADIUM COURSE. One of America's most challenging courses. 56 min., $17.95.

ST. ANDREWS OLD COURSE. The birthplace of golf. 60 min., $17.95.

TPC AT SAWGRASS. Experience this legendary course. 35 min., $17.95.

LEE TREVINO'S LEGACY OF THE LINKS. Trevino narrates the history of the hallowed St. Andrews golf course. (PAR, 1987) 90 min., $19.95.

TOUGHEST 18 HOLES OF GOLF IN AMERICA. Hosted by Gary

Player; each hole is played by a club or touring pro. 76 min., $49.95.

WALT DISNEY WORLD: MAGNOLIA. The course's best-kept secrets. 45 min., $17.95.

Individual Profiles

Seems like there should be more of these. Maybe great golfers just don't lead as interesting lives as, say, matadors. The Greatest Sports Legend Series is now called Video Card for most sports.

SEVE BALLESTEROS: *A Study of a Legend*. Profiles the Spanish golf master. $19.99.

TOM MORRIS: KEEPER OF THE GREENS. A docudrama about the grand old man of golf and the game in the 1860s. (1992) 63 min., $22.95.

ARNOLD PALMER. Greatest Sports Legends. (RVI) 30 min., $6.98.

ARNOLD PALMER STORY. Golf's first superstar. (1992) 55 min., $22.95.

SAM SNEAD. Greatest Sports Legends. (RVI) 30 min., $6.98.

LEE TREVINO. Greatest Sports Legends. (RVI) 30 min., $6.98.

Instructionals

Even if you have been told that the only thing that can cure your game is a trip to Lourdes, you're sure to find some helpful tips among these videos. Don't think of it as a lesson on your TV; think of it as having Lee Trevino as your personal pro. Or Chi Chi Rodriguez. Or Jack Nicklaus, but he'll cost you about four times what the others will.

Men

PETER ALLISS: PLAY BETTER GOLF. Personal coaching from tee to putt. 116 min., $29.95.

TOMMY ARMOUR: HOW TO PLAY YOUR BEST GOLF ALL THE TIME. All aspects of the game. 38 min. B&W with 151-page workbook.

WALLY ARMSTRONG: A PICTURE'S WORTH 1000 WORDS. Builds a picture of proper fundamentals through creative props. 30 min., $19.95.

WALLY ARMSTRONG: FEEL YOUR WAY TO BETTER GOLF. Teaches the feel of a powerful swing. (SIM) 52 min. $9.95.

WALLY ARMSTRONG: GOLF GADGETS & GIMMICKS. Unique gadgets and training aids to help you improve. 30 min., $16.95.

WALLY ARMSTRONG'S GOLF: THE EASY WAY. How to take off strokes in 30 minutes. 30 min., $19.95.

WALLY ARMSTRONG'S COLLECTION OF TEACHING AIDS AND DRILLS. Sold only as a set. 32 min. each, $84.95 for set.

    GATOR GOLF DRILLS: VOL. 1. Alignment and stance, path and plane, grip and full swing.

    GATOR GOLF DRILLS: VOL. 2. Back swing, forward swing, full swing, balance drills, upper body drill, tempo and timing.

    SHORT GAME TEACHING DRILLS. Putting, chipping, pitching, and sand shots.

    SHORT GAME TEACHING AIDS. 67 unusual short game aids.

    TEACHING AIDS, VOL. 1. 37 different aids to teach alignment and stance, path and plane, grip and full swing.

    TEACHING AIDS, VOL. 2. 41 training aids to teach full swing, back swing, forward swing, leg and footwork.

THE AZINGER WAY. "All the things I've learned," says Paul Azinger. 55 min., $24.95.

JIMMY BALLARD: FUNDAMENTAL GOLF SWING. Teaches the seven common denominators of a good swing. 69 min., $49.50.

BIOMECHANICS OF POWER GOLF. How to create more energy transfer. Includes two audiocassettes and 29-page booklet. 60 min., $59.50.

BRAINWAVES GOLF. Dr. JoAnne Whitaker teaches the mental side. 30 min., $19.95.

BILLY CASPER: GOLF BASICS. (SIM) 30 min., $17.95 each.

   VOL. 1. Grip, stance, swing, driving, and more.

   VOL. 2. Hooks, slices, long putts, difficult lies, more.

BILLY CASPER: GOLF LIKE A PRO. One of golf's all-time greats shares his knowledge. (MOR) 50 min., $19.95.

BILLY CASPER: SECRETS OF GOLF. Casper provides winning tips on club selection, sand play, playing trouble shots, etc. (MOR) 30 min., $14.95.

BOB CHARLES: GOLF FROM THE OTHER SIDE. Instruction for lefthanders. 60 min. $49.95.

CHI CHI'S BAG OF TRICKS: *In and Out of Trouble with Chi Chi Rodriguez.* How to get out of heavy rough, out from under trees, etc. (CBS, 1988) 58 min., $49.98.

THE CHIPPING AND PUTTING VIDEO. Charlie Schnauble shows the quickest way to improve. 30 min., $14.95.

BRUCE CRAMPTON: TOTAL GOLF. The twenty-year PGA veteran shares his secrets. 92 min., $29.95.

BEN CRENSHAW: THE ART OF PUTTING. Covers grip, stance, stroke, reading greens, etc. (HPG, 1986) 44 min., $34.95.

MIKE DUNAWAY: POWER DRIVING. Techniques of "the world's longest driver." (SYB) 30 min.,, $49.95.

CHARLIE EARP'S GOLF LESSON. Australia's master coach with a wealth of tips. 30 min., $24.95.

ETIQUETTE OF GOLF. Proper etiquette from tee to cup. (1992) 30 min., $19.95.

NICK FALDO IS GOLF. Secrets of golfing from the British Open Champion. (VES, 1989) 60 min., $29.98.

NICK FALDO'S GOLF COURSE. Faldo's secrets to golf success. $29.98.

FANTASTIC APPROACHES: *The Pro's Edge*. Compendium of short-iron finesse featuring top pros. (PAR, 1990) 38 min., $19.95.

RAY FLOYD: 60 YARDS IN. Comprehensive video dedicated to the short game. 60 min., $39.95.

THE FOUR ABSOLUTES. All great golf swings have four distinct elements in common. 65 min., $39.95.

AL GEIBERGER: 6-IN-1 GOLF CLINIC. Six golf clinics in one. 60 min., $24.95.

AL GEIBERGER: WINNING GOLF. Five easy-to-follow lessons. 60 min., $24.95.

AL GEIBERGER: GOLF. Groove the image and reach deep into your mind with the SyberVision muscle-memory training technique. (SYB) 60 min. video, 4 audiotapes, training guide, $84.95. Video only $69.95.

GOLF DIGEST VIDEOS. (GD) Each is 26 min., $29.50.

- VOL. I: A SWING FOR A LIFETIME. Bob Toski and Jim Flick. The three main components of a successful swing.
- VOL. 2: FIND YOUR OWN FUNDAMENTALS. Bob Toski and Jim Flick. How to adapt five key preswing fundamentals.
- VOL. 3: DRIVE FOR DISTANCE. Bob Toski and John Elliott. Four secrets to longer tee shots.
- VOL. 4: SHARPEN YOUR SHORT IRONS. Bob Toski and Jim Flick. Four fundamentals to better short iron play.
- VOL. 5: SAVE FROM THE SAND. John Elliott. The main elements that affect successful bunker play.
- VOL. 6: PUTTING FOR PROFIT. Tom Ness. Comprehensive guide to putting.

VOL. 7: WHEN THE CHIPS ARE DOWN. Bob Toski and Jack Lumpkin. Learn the two main chipping motions.

VOL. 8: WINNING PITCH SHOTS. Davis Love, Jr. The shot you need, the club, how to shape the shot, proper execution.

VOL. 9: HITTING THE LONG SHOTS. Davis Love, Jr. Take the fear out of long irons and fairway woods.

VOL. 10: TROUBLE SHOTS; GREAT ESCAPES. Hank Johnson and Bob Toski. Minimize damage.

GOLF FOR WINNERS. Mark O'Meara and Hank Haney show how to play "winning" golf. 42 min., $22.95.

GOLF FUNDAMENTALS: *The Ben Sutton Golf School.* Designed to improve play without abandoning the natural swing. (ESP, 1986) 60 min., $29.95.

GOLF TIPS FROM 27 TOP PROS. Snead, Boros, Rosburg, Palmer, and 23 others display their techniques. (PAR, 1989) 40 min., $19.95.

GOLF WITH THE SUPER PROS. Hear from Snead, Palmer, Trevino, Irwin, etc. (1992) 42 min., $19.95.

JACK GROUT: KEYS TO CONSISTENCY. Focuses on elements to establish consistency. 60 min., $39.95.

JACK GROUT: THE LAST 100 YARDS. Techniques for playing the last 100 yards of the game. 45 min., $39.95.

JOHN HAINES: GOLF ETIQUETTE. Everything you need to know from proper attire to attitude. (1992) 35 min., $29.95.

JACK HAMM: HIT IT LONG. The holder of the Guinness World Record for the Longest Drive in a PGA event (437 yds) shows how. (1992) 30 min. $24.95.

HOW TO BREAK 90 IN 30 DAYS. Bob Kurtz and daughter show ten tips for lower scores. (BFS) 50 min., $19.50.

AN INSIDE LOOK AT THE GAME FOR A LIFETIME. Bob Toski, Jim Flick, Peter Kostis, and John Elliott of the *Golf Digest* School. (GD) 56 min., $49.95.

HALE IRWIN: DIFFICULT SHOTS. The two-time U.S. Open champ

demonstrates how to tackle the toughest shots. (SYB) 60 min., $49.95.

JOHN JACOBS: FULL SWING. Author of *Practical Golf* explains the full swing from tee to green. (GD) 57 min., $69.50.

JOHN JACOBS: SHORT GAME. Putting and shots around the green. (GD) 60 min., $69.50.

JOHN JACOBS: FAULTS AND CURES. 18 simple and practical ways to better golf. (GD) 58 min., $69.50.

BOBBY JONES LIMITED COLLECTOR'S EDITION. Two instructional tapes utilizing recently discovered film to demonstrate Jones's complete game. Also includes the book *A Golf Story* (Charles Price's remembrance of Jones) and certificate of authentication. (SYB, 1992) 90 min. each, $245.00.

THE BEST OF BOBBY JONES. Jones's favorite six instructional tapes. (SYB) 70 min. video, 3 USGA-approved signature golf balls and 258-page biography. $99.85.

BOBBY JONES INSTRUCTIONAL SERIES. Highlights the essentials of Jones's game. (SYB, 1992) 60 min. each.

  THE FULL SWING. $69.95.

  FROM TEE TO GREEN. $69.95.

KILLER GOLF: *10 Tips to Take 10 Strokes Off Your Game.* With Gary McCord. (1992) 40 min., $14.98.

TOM KITE: REACHING YOUR GOLF POTENTIAL.

  VOL. 1: DEVELOPING MAXIMUM CONSISTENCY. The 1989 PGA Golfer of the Year takes you through practice sessions from tee to green. 50 min., $28.50.

  VOL. 2: STRATEGIES & TECHNIQUES. Kite analyzes various situations in a round. 60 min., $28.50.

GEORGE KNUDSON: THE SWING MOTION. Simple techniques for golfers at any level. 25 min., $24.95.

DAVID LEADBETTER: THE FULL GOLF SWING. Step-by-step by the "World's Number One Coach." 90 min., $59.95.

DAVID LEADBETTER: THE SHORT GAME. Shows why the key to lower scores is the short game. 100 min., $59.95.

BILL LINTON: BETTER GOLF WITH A LITTLE BIT OF MAGIC. Sound golfing techniques with visual aids. 45 min., $29.95.

GENE LITTLER: THE 10 BASICS. Home practice-play program. 45 min., $19.95.

BOB MANN'S COMPLETE AUTOMATIC GOLF METHOD. Combines two earlier, shorter videos plus instruction in specialty shots. (VID, 1992) 80 min., $19.98.

See also:

BOB MANN'S AUTOMATIC GOLF: *The Method.* Contains two videos. (VID) 44 min., $9.95.

BOB MANN'S GOLF: *The Specialty Shots.* (VID) 30 min., $9.95.

BOB MANN: *Let's Get Started.* A swing you don't have to think about. (VID, 1992) 45 min., $12.95.

THE MASTER SYSTEM TO BETTER GOLF

VOL. I. Davis Love III on driving, Tom Purtzer on iron accuracy, Craig Stadler on the short game, and Gary Koch on putting. (FH, 1987) 60 min., $19.95.

VOL. II. Paul Azinger on fairway and green sand traps, Fred Couples on tempo, and Bobby Wadkins on picking the right club to get out of trouble. (FH, 1988) 60 min., $19.95.

Also see at $9.95 each:

PAUL AZINGER ON FAIRWAY AND GREEN SAND TRAPS. (FH, 1988) 20 min.

FRED COUPLES ON TEMPO. (FH, 1988) 20 min.

GARY KOCH ON PUTTING. (FH, 1987) 20 min.

DAVIS LOVE III ON DRIVING. (FH, 1987) 20 min.

TOM PURTZER ON IRON ACCURACY. (FH, 1987) 20 min.

CRAIG STADLER ON THE SHORT GAME. (FH, 1987) 20 min.

BOBBY WADKINS ON TROUBLE SHOTS. (FH, 1988) 20 min.

MICHAEL McTEIGUE: KEYS TO THE EFFORTLESS GOLF SWING. Eight easy lessons to build muscle memory. 80 min., $39.95.

JOHNNY MILLER: GOLF THE MILLER WAY. Tips and tricks for golfing success. (MOR) 30 min., $29.95.

KELLY MURRAY: POWER DRIVING. The Canadian shows how he became International Long Drive Champion. 60 min., $22.95.

NICE SHOT! Installs a perfect golf swing in your mind. 65 min., $39.95. (Also available with two audiotapes and workbook for $69.95.)

JACK NICKLAUS' GOLF MY WAY I: *Hitting the Shots.* More than two hours of personal instructions by golf's master. Winner of numerous video awards. (WV, 1988) 128 min., $84.95.

JACK NICKLAUS' GOLF MY WAY II: *Playing the Game.* Emphasizes strategy and tactics with 60 easy-to-follow lessons. (WV, 1990) 141 min., $85.00.

JACK NICKLAUS' GOLF MY WAY: *The Full Swing.* Abridged version of "Golf My Way" videos. (GD) 38 min., $19.95.

9 TIPS FROM 9 LEGENDS OF GOLF

> VOL. I: FROM TEE TO FAIRWAY. Sam Snead, Miller Barber, Mike Souchak, Peter Thompson, Butch Baird, Doug Ford, Art Wall, Tommy Bolt, Gene Littler. (1992) 60 min., $16.95.

> VOL. 2: FROM FAIRWAY TO GREEN. Don January, Gardner Dickinson, Doug Sanders, Jerry Barber, Julius Boros, Charlie Sifford, Gay Brewer, Billy Casper, Bob Goalby. (1992) 60 min., $16.95.

GREG NORMAN: COMPLETE GOLFER

> PART 1. "The Great White Shark" shows you how to crush the ball. (PAR, 1986) 63 min., $29.95.

> PART 2. Norman demonstrates chipping and putting. (PAR, 1989) 46 min., $29.95.

ONE MOVE TO BETTER GOLF. Teaching pro Carl Lohren shows his "one move" program to improve your game. (BFV) 30 min., $19.95.

ARNOLD PALMER: PLAY GREAT GOLF. An oldie but a goodie.

> VOL. 1: MASTERING THE FUNDAMENTALS. (VES) 60 min., $39.98.
>
> VOL. 2: COURSE STRATEGY. (VES) 60 min., $39.98.
>
> VOL. 3: THE SCORING ZONE. (VES) 60 min., $39.98.
>
> VOL. 4: PRACTICE LIKE A PRO. (VES) 60 min., $39.98.

PGA TOUR GOLF. Top stars of the PGA Tour lend their combined expertise in this three-video series.

> PGA TOUR GOLF 1: *The Full Swing*. Correct stances, grips, and follow-throughs. (1988) 60 min., $19.95.
>
> PGA TOUR GOLF 2: *The Short Game*. Fundamental chipping and putting tips. (1988) 60 min., $19.95.
>
> PGA TOUR GOLF 3: *Course Strategy*. Reading a course, choosing among options, and strategy. (1988) 60 min., $19.95.

PLAY YOUR BEST GOLF. Bob Toski, Jim Flick, Gary Wiren, Peggy Kirk Bell, et al.

> VOL. I: THE CLUBS. From preswing fundamentals to course strategy with wood and iron. 69 min., $26.95.
>
> VOL. 2: THE STRATEGIES. Approach shots, chips, putts, and developing confidence. 109 min., $26.95.

also

> APPROACH & SAND PLAY. Trouble shots, bunker lies, four levels of approach play. 17 min., $19.95.
>
> GOLF SWING. Ball flight laws, building a swing. 20 min., $19.95.
>
> MID IRONS & SHORT IRONS. Grip, alignment, swing, club selection, and more. 13 min., $19.95.
>
> PUTTING & CHIPPING. Grip, stance, body motion, etc. 20 min., $19.95.
>
> STRATEGIES & SKILLS. Stretching, driving, putting tips, and more. 25 min., $19.95.
>
> WOODS & LONG IRONS. Tee height, stance, etc. 19 min., $19.95.

GARY PLAYER ON GOLF. Player shares his personal tips for developing the "natural" feel. 90 min., $29.95.

PHIL RITSON: GOLF YOUR WAY. 39 specific drills to improve your game. 78 min., $22.95.

PHIL RITSON: VIDEO ENCYCLOPEDIA OF GOLF. Ritson discusses elements of the game in these programs. 30 min., $19.95 each.

- VOL. 1: GRIP, STANCE, AIM & POSTURE
- VOL. 2: CONTROL YOUR SWING PLANE
- VOL. 3: CURES FOR CROOKED SHOTS
- VOL. 4: GOLF'S MENTAL KEYS
- VOL. 5: EFFORTLESS POWER
- VOL. 6: SECRETS OF THE POWER FADE
- VOL. 7: PITCHING, CHIPPING, PUTTING
- VOL. 8: SAND MAGIC
- VOL. 9: LADIES, EFFORTLESS POWER
- VOL. 10: SENIORS, EFFORTLESS POWER
- VOL. 11: USING THE WIND TO WIN

PHIL ROGERS: THE SHORT GAME. The "ins" and "outs" of the short game. 72 min., $39.95.

PUTTING WITH CONFIDENCE. Duff Lawrence and Barb Thomas show how confident putting can lower your score. (BFV) 30 min., $24.95.

BOB ROSBURG'S BREAK 90 IN 21 DAYS. Step by step. 35 min., $9.95.

BOB ROSBURG: GOLF TIPS. Tips on playing the short game. 60 min., $59.95.

RULES OF GOLF. Tom Watson, Juli Inkster and Peter Alliss explain. 30 min., $9.95. (Available with USGA Rule Book for $5 more through some dealers.)

PAUL RUNYON: THE SHORT WAY TO LOWER SCORING. Golf

Digest's teaching pro. (GD)

    VOL. I: PUTTING & CHIPPING. 35 min., $29.50.

    VOL. II: PITCHING & SAND PLAY. 27 min., $29.95.

SAVING STROKES WITH THE RULES. Use the rules to help you play better and faster. 32 min., $14.95.

JOHN SCHLEE: MASTERING THE LONG PUTTER. A video dedicated solely to perfecting the long putter. 50 min., $29.95.

JOHN SCHLEE: MAXIMUM GOLF. Comprehensive instructional course designed by John Schlee. Includes 3-hour audiocassette and 148-page booklet. 120 min., $49.95.

ART SELLINGER: DRIVE MY WAY. One of the world's longest drivers shows you how. 30 min., $19.95.

DAVE STOCKTON: PRECISION PUTTING. Two-time PGA champ shows how to master the shortest shot. (SYB) 30 min., $49.95.

DAVE STOCKTON'S GOLF CLINIC. Learn basics. (SYB) 60 min., $49.95.

CURTIS STRANGE: HOW TO WIN AND WIN AGAIN. Strange teaches techniques from fundamentals to fine points. (NAC, 1990) 70 min., $29.98.

10 FUNDAMENTALS OF THE MODERN GOLF SWING. David Glenz and Jim McClean teach ten critical keys. 32 min., $22.95.

TIGER SHARK VIDEO LIBRARY. With Pat Simmons. (1990) 15 min., $19.95 each.

    IRONS: *Hitting with Power and Accuracy.*

    PUTTING: *The Science and the Stroke.*

    TROUBLE SHOTS: *Recovery Strkes Made Easy.*

    WOODS: *Hitting for Distance.*

TIPS FROM THE TOUR, VOLUME 1. Collection of tips and drills by eleven PGA pros and three senior pros. (IVE, 1988) 35 min., $14.95.

BOB TOSKI TEACHES YOU GOLF. The Dean of the *Golf Digest*

Instruction School demonstrates all phases. (GD) 70 min., $48.95.

LEE TREVINO'S PRICELESS GOLF TIPS

VOL. 1. Improving your approach shots and putting. (PAR, 1987) 25 min., $19.95.

VOL. 2. How to get out of sand, bunkers, and tricky lies. (PAR, 1987) 27 min., $19.95.

VOL. 3. Gaining distance and accuracy with your drives. (PAR, 1987) 25 min., $19.95.

LEE TREVINO'S PUTT FOR DOUGH. Choosing a putter, reading the green, avoiding "yips," etc. (PAR, 1989) 50 min., $24.95.

25 GREAT PROS' SECOND SHOTS. Littler, Bolt, Palmer, and others use their second shots to help them break par. (PAR, 1990) 36 min., $19.95.

KEN VENTURI: BETTER GOLF NOW! Fundamentals from tee to green with drills and memory aids. (HPG, 1987) 40 min., $39.95.

KEN VENTURI: STROKE SAVERS. Get out of trouble with Venturi. 60 min., $22.95.

TOM WEISKOPF: THE GOLF SWING. The basics of grip, stance, posture and balance for tee and fairway. 45 min., $19.95.

DR. GARY WIREN: THE GREATER GOLFER IN YOU.

VOL. 1. Includes a repeating golf swing, unleashing power, mental tips, more. 84 min., $31.95.

VOL. 2. Includes swing imagery, pitching, chipping, putting, bunker play, more. 87 min., $31.95.

IAN WOOSNAM: POWER GAME. The 1991 Masters champ demonstrates his relaxed yet powerful swing. (1992) 55 min., $19.95.

KERMIT ZARLEY: GOLF FOR ALL AGE GROUPS. Technique, strategy, tactics, and "mental imaging." 25 min., $19.95.

FUZZY ZOELLER: SCRAMBLE TO BETTER GOLF. How to recover from bad shots. 42 min., $18.95.

Women

AMY ALCOTT: WINNING AT GOLF. A proven winner helps you learn from tee to pin. (1992) 37 min., $19.50.

JOANNE CARNER'S KEYS TO GREAT GOLF. Straighter drives and more accuracy with irons can cut ten strokes off your game. 90 min., $39.95.

BETSY CULLEN: HIT IT FARTHER. The easy system that improves distance. 60 min., $29.95.

NANCY LOPEZ: GOLF MADE EASY. One of the top woman golfers shows how. (FH) 48 min., $19.98.

MASTERING THE BASICS. A guide for the woman golfer for both beginners and accomplished players. (WKV, 1990) 40 min., $19.95.

JUDY RANKIN'S BREAK 90 IN 21 DAYS. Step-by-step instruction for women golfers. 35 min., $9.95.

PATTY SHEEHAN: SYBERVISION WOMEN'S GOLF. From drive to putt. 60 min., $69.95. With 4 audiocassettes and training guide, $84.95.

JAN STEPHENSON'S HOW TO GOLF. Instructions on every aspect of the game. (WA) 50 min., $29.95.

DONNA WHITE'S BEGINNING GOLF FOR WOMEN. (SIM) 40 min., $19.95.

DONNA WHITE: GOLF FOR WOMEN. Correct grip, posture and alignment, swing fundamentals, practice tips, putting. 30 min., $14.95.

WOMAN'S GOLF FUNDAMENTALS: *The Ben Sutton Golf School.* Joyce Ann Jackson offers friendly and expert instruction. (ESP, 1988) 60 min., $29.95.

WOMEN'S GOLF INSTRUCTIONAL SERIES. Peggy Kirk Bell, Dede Owens, Annette Thompson, Carolyn Hill, and Muffin Spencer-Devlin. 40 min., $24.95 each.

    VOL. 1: THE FULL SWING

VOL. 2: THE APPROACH GAME

Seniors

JAY OVERTON: 50 PLUS SENIOR GOLF. Helping the 50 and over golfer maximize his performance. 60 min., $39.95.

PLAY SENIOR GOLF. Littler, Nelson, Boros, Casper, and others give their tips. (BFV) 72 min. $19.95.

SENIORS. Stars of the Senior PGA Tour; Miller Barber on the driver and the wedge, Orville Moody on long irons and putting, and Dale Douglass on rhythm, tempo, sand, and chip shots. (FH, 1988) 115 min., $39.95.

See also

> MILLER BARBER ON THE DRIVER AND WEDGE. (FH, 1988) 25 min., $9.95.
>
> DALE DOUGLASS: RHYTHM, TEMPO, SAND, AND CHIP SHOTS. (FH, 1988) 25 min., $9.95.
>
> ORVILLE MOODY: LONG IRONS AND PUTTING. (FH, 1988) 25 min., $9.95.

SAM SNEAD'S SECRETS FOR SENIORS. Slammin' Sam shares his shotmaking secrets and his approach to physical and mental fitness. 58 min., $29.98.

Juniors

WALLY ARMSTRONG'S GOLF FOR KIDS OF ALL AGES. 40 min., $19.95.

SCHOLASTIC GOLF. Eric Alpenfels and PGA pro Jay Haas teach drills and instructions for the young game. 47 min., $19.95.

SUPER GOLF FOR JUNIORS. Dr. Joan Churchill teaches managing golf stress and pro Mike Barge shows how to correct problems. 45 min., $19.95.

TEACHING KIDS GOLF. Experts from the famous Ben Sutton Golf School provide instruction for children and teachers of children. (ESP, 1987) 60 min., $29.95.

LEE TREVINO'S GOLF TIPS FOR YOUNGSTERS. Trevino uses his trademark humor to instruct kids in basics. (PAR,1988) 40 min., $19.95.

Wide World of Golf: Video Magazine

Each video is 60 min., $29.95. (We found no listing for Vol. 1: Issue 3.)

VOL. 1: ISSUE 1. The grip with Nicklaus, Norman, and Bobby Jones. Chipping with Tom Kite. Spanish Bay Resort. Curtis Strange profile. The Golden Bear remembers.

VOL. 1: ISSUE 2. Keys to accuracy with Norman, Nicklaus, and Trevino. Swing test with Faldo, Nicklaus, and Weiskopf. Florida's Grand Cypress. Masters preview. Watching the LPGA.

VOL. 1: ISSUE 4. British Open review. Golfing in Southwest Ireland. Wedge tips. Payne Stewart profile. Nicklaus and Strange on head movement. Pitching over bunkers. Byron Nelson's swing analyzed.

VOL. 1: ISSUE 5. The longest ball. Irwin's winning technique. Putting guide. Robert Gamez profile. Palmer on uneven lies. Golfing in Palm Springs. Judy Rankin's alignment tips. A look at Snead's swing.

VOL. 1: ISSUE 6. Simpson, Zoeller and King on putting. Swing plane tips. Celebrity lessons. Palmer on water. Backswing tips. Nicklaus, Jones and Strange on transition. Dan Jenkins profile.

VOL. 2: ISSUE 1. "How to nail it." PGA advice. Fuzzy Zoeller's video. Total Putting Guide: Part 3. Releasing the clubhead. Carrying a third wedge. Harry Vardon's swing. Celebrity lessons with Mike Schmidt. Nicklaus' first year.

VOL. 2: ISSUE 2. An inside look at Augusta National. Ten Most Common Faults. Profile: Nick Faldo. Tips from Chi Chi. Using the rules. Rankin on the Practice Tee. Putting Guide: Series Conclusion. Chipping.

VOL. 2: ISSUE 3. U.S. Open at Hazeltine. Irwin analyzes Palmer's swing. Common faults. Jack Grout on watching the ball. Robert Trent Jones. Nicklaus & Palmer on the rough. Azinger's advice. Using the rules.

VOL. 2: ISSUE 4. Highlights from the '91 Masters. Stop choking. Faldo's technique. Profile: Phil Mickelson. Mid-round trips. Stop hitting thin. Rankin on relaxation. Using the rules. Palmer & Norman on the fade.

VOL. 2: ISSUE 5. 1991 U.S. Open highlights. Short game with Payne Stewart. Pressure putts. Mid-round adjustments. Using the rules. Peter Alliss looks at the Ryder Cup.

Fitness and Conditioning

EXERCISE FITNESS FOR GOLF. Exercise programs specifically for golf. Researched by Dr. Art Dickson, featuring Dow Finsterwald. 60 min., $29.95.

EXERCISE FITNESS FOR GOLF SERIES. 30 min., $16.95 each.

> VOL. 1. Upper and lower body stretching.
>
> VOL. 2. Manual resistance exercises for strength and endurance.
>
> VOL. 3. Two-weight exercises for more advanced muscle training.

EXERCISES FOR BETTER GOLF. Dr. Frank Jobe teaches exercises to strengthen golf muscles. 73 min., $29.95.

FIT FOR GOLF. Dr. Marnin Clein presents an exercise program to increase muscle strength. 30 min., $19.95.

GOLFLEX. John Elliott presents the Golflex System of Exercises. (1992) 50 min., $29.95.

JOHN RHODES: GET FIT FOR GOLF. Training "golf specific" muscles to reduce fatigue and injuries. With Exertube, an exercise aid. 60 min., $29.95.

SPORTS PSYCHOLOGY: *The Winning Edge in Sports*. Improving mental performance. 60 min. plus audio cassette, $44.95.

STRETCHING: *The Driving Force*. Easy to follow pregolf stretching routine. 19 min., $16.95.

STRETCHING FOR BETTER GOLF. Stretching exercises to prevent injuries and lower your score, with Bob Anderson. 40 min., $19.95.

DR. GARY WIREN: SUPER POWER GOLF. Three-part program that combines increased strength, flexibility, and swing techniques. 55 min., $39.95.

Bloopers and Humor

"Hit the ball and pull Henry; hit the ball and pull Henry!" If you laughed at that one oh so many years ago (or, if you even remember the setup lines), you're probably able to laugh at the game, at least a little. Here are some chuckles on video.

ACE OF CLUBS. Paul Hahn, Jr. performs some of his most famous trick shots. 20 min., $16.95.

BIRDIES AND BLOOPERS. Some of golf's funniest moments by some of its great stars. 30 min., $9.95.

DORF ON GOLF. Comic Tim Conway teaches golf the "Derk Dorf

Way." (J2) 30 min., $19.95.

DORF'S GOLF BIBLE. Heee's baaack! The hilarious Conway character even takes a lesson from Sam Snead in this one. (J2) 36 min., $19.95.

GOLFBUSTERS: *The Lighter Side of Golf.* Combines comedy with golf trivia and history. 40 min., $18.95.

GOLF COMEDY CLASSICS. Classic golf routines from TV's Jackie Gleason and Lucille Ball. 60 min., $14.95.

GOLF FUNNIES. Weird and wacky shots by pros and celebrities. (1987) 30 min., $9.95.

GOLF TRICK SHOTS. With Paul Hahn, Jr. 20 min., $16.95.

GOLF'S GAMBLING GAMES. A thorough yet humorous look at the games golfers play. (1992) 55 min., $24.95.

GOLF'S GOOF-UPS & MIRACULOUS MOMENTS. Some of golf's wildest moments, including awesome putts, crazy shots. (1992) 45 min., $19.95.

GOOD GRIEF! GOLF?. Strange, remarkable and fantastic shots. 38 min., $14.95. (Also available in a comedy 2-pack with *One Club Challenge* for $22.95.)

HOW TO WIN AT GOLF WITHOUT REALLY CHEATING. A humorous look at 27 winning golf ploys. (1992) 35 min., $19.95.

JUST MISSED, DAMMIT! Some of golf's greatest players blow tap-ins, slice drives, and play like duffers. (PAR, 1988) 40 min., $19.95.

LITTLE SCAMS ON GOLF. Rich Little and 18 of his "friends" do 18 holes. 44 min., $17.95.

MURPHY'S LAWS OF GOLF. If it can go wrong, it will for Tom Posten. 30 min., $24.95.

ONE CLUB CHALLENGE. Trevino and Ballesteros challenge Faldo and Aoki to a one-club match at St. Andrews. 42 min., $14.95. (Also available in a comedy 2-pack with *Good Grief! Golf?* for $22.95.)

OUTRAGEOUSLY FUNNY GOLF. Trick shot artist Paul Hahn with irregular props. 30 min., $19.95.

THOM SHARP'S GOLF: I HATE THIS GAME! The comedian provides humorous tips on attire, cheating, heckling opponents, and other fine points of the game. (PAR, 1988) 30 min., $19.95.

THREE MEN AND A BOGEY. Humor. 30 min., $9.95.

WORLD'S WORST AVID GOLFER. Compilation of some of the worst rounds ever played, plus cart crashes and Ms. Dixie Cupps, the world's sexiest golfer. 30 min., $9.98.

See the note in the Introduction concerning Golf Videos USA.

GENERAL, OLYMPICS, AND OTHER SPORTS

The title of this chapter may raise a few hackles. What, you may ask, is a "general" sport? And how can we lump your favorite sport (bowling?) in with all those "others"? And what about the Olympics? Surely the Olympics deserve a chapter of their own. Maybe two!

To all those with real or fancied wounds, we apologize and offer the lame excuse that none of the sports within this chapter—not even the Olympics—offers enough videos to justify a whole chapter of its own. Skiing came close.

"General" Sports Videos

Now, as to "general" sports videos. These are conglomerates of highlights or lowlights from several different sports. A highlight video such as *Great Sports Moments of the '80s*, for example, offers the variety of the morning newspaper's sportspage, all dished up with an Al Michaels narration. You can watch Kirk Gibson hit his gimpy-legged home run off Dennis Eckersley, Michael Jordan fly through the air with the greatest of ease, Eric Heiden approach jet speed on skates, Doug Flutie complete an Orange-Bowl-winning pass, Carl Lewis in two Olympics, and native-Augustan Larry Mize win the 1987 Masters in sudden-death. Or California's hot-potato act with the football to beat Stanford, Mike Schmidt's 500th home run, Ray Leonard's split-decision victory over Marvin Hagler, or goalie Ron Hextall scoring a goal.

The format is to pick in descending order the top ten sports moments of the decade. (A hint: Number One isn't mentioned above.) That invites controversy; just try to find two people who'd pick the same ten moments in the same order. You'll more easily find two guys with the same telephone number. However, that aside, the strength of this video—there's something for everyone—is also its weakness—there's not enough of anything. Each of the top ten moments gets about a minute and a half of tape time, not really enough to build dramatic intensity. Sprinkled around each of the major events are a half-dozen or so fifteen-second clips of other moments that cry out for extension.

On the other hand, when the limelight shines on a sport that doesn't tweak your interest, the moment is over so quickly it's hardly worth fast-forwarding.

The primary appeal of this kind of video is in the sports that aren't "important enough" to have their own tapes. But, in truth, most of the magic moments *can* be found elsewhere. Instead of a minute and a half of the '88 World Series, why not go out and get the whole thing?

Most of the general sports videos are built around a time period, like the 1980s, or a place. You might enjoy *The Greatest Moments in Chicago Sports History* more if you're from Cook County, but most of the moments made national headlines.

A few videos trade on personalities. The Greatest Sports Legends serves up anywhere from 16 to 160 superstars on different tapes.

Blooper videos are about as common as highlights. Tim McCarver on *The Not-So-Great Moments in Sports* is perhaps the best known, but Bob Uecker, Roy Firestone, and others have tried to raise chuckles with various inventories of pratfalls, gaffes, and goofs. Although these and others are aimed at adults (who are likely to buy them), they would seem to be ideal for kids who get to see some sports they might otherwise ignore.

Naturally, there's no such animal as a general sports instructional. If you can imagine how to putt, how to pole vault, and how to run a fly pattern on the same video, you'll understand why.

Olympic Videos

If you remember, the Ancient Greeks were among the most quarrelsome people ever. When they weren't fighting Persians, they were busy trying to dismantle each other. Athens fought Sparta. Thebes fought Corinth. Sparta fought Thebes. And on and on, until the Romans finally moved in and took over. But before the *Pax Romana*, the Greeks held this big sports tournament every four years at Mt. Olympus. How big was it? So big that while it was going on, all the wars were put on hold.

It was the with this pacifying idea in mind that the Baron Pierre de Coubertin of France thought up the idea of the modern Olympics in 1896. If people from all over the world could compete on the playing field, they might abstain from competing on the battlefield. Sad to say, the Baron's dream hasn't quite worked out, which is one of the reasons no Olympics were held in 1916, 1940, or 1944, when war took precedence over games.

Even sadder is the politicizing of the Olympics. We've seen Hitler's attempt to promote a Master Race in Berlin, Black Power salutes in Mexico, terrorism in Munich, and abstentions for political reasons by both the U.S.A. and the U.S.S.R. The regular totaling of gold, silver, and bronze medals by country is another sad note. Us vs. them wasn't supposed to be part of it.

Another thing that would set the Baron spinning in his grave is the commercialization of the event. We have "official" everything from breakfast food to luggage. Want to sell your widgets as the "Official Widget of the Olympics?" Just contribute enough to the IOC.

Some of the lengths companies go to would be laughable if they didn't break your heart. At this writing, the following is an "Olympic" story. Visa bought the right to be the official plastic for the 1992 Winter Olympics and promptly crowed in their ads that the folks there wouldn't "take" American Express. Am-Ex countered with ads showing skiers and such using *their* cards amid the snow of the Alps, suggesting that there were some folks over there who just wanted to get paid, no matter which brand of credit card was used. Visa then sued American Express on the grounds apparently that being official meant they'd purchased an option on the French Alps for the winter of '92. Just how this will all turn out, we don't know or particularly care, but none of it seems to have much to do with who can skate, ski, or luge best.

By the time you read this, the market will be flooded with videos of the Winter and Summer Olympics. The number and length will probably depend on how well the U.S. does medalwise.

Past glories are available on such as *The Official* (there's that word again!) *1988 Calgary Winter Olympic Video* or *Time Capsule: Los Angeles Olympic Games of 1932*. And, for all the crap, every Olympic competition has supreme moments well worth remembering.

We were unable to find a Greatest Olympics Bloopers tape. Somebody out there is missing a good bet.

Other Sports

We know you'd like to see all the videos relating to your favorite sport together in one section by themselves, but the reason we lumped bowling, skiing, horse racing, and a couple of others together is your fault. You just haven't been buying enough of these videos to get producers to inundate us with titles.

Obviously, certain sports lend themselves to certain kinds of videos. Any video on horse racing is sure to be a highlight type. When a horse takes a fall, it's not very funny. And an instructional video would be wasted on any steed save Mr. Ed. The bowling video field is entirely instructional. Great sport, but one strike looks pretty much like any other. Skiing, however, touches all the bases—highlights, instructionals, and bloopers.

Skiing videos are fine to watch, but, will any of them get me on the slopes myself? Personally, I'll take my shushing at the library. I'm guided by a statement newswoman Linda Ellerbee attributes to her husband: "Any sport in which the Red Cross plays a significant role probably [is] not right for a naturally clumsy person."

General, Olympics, and Other Sports Videos

Highlights and History

MARIO ANDRETTI. Greatest Sports Legends. (RVI) 30 min., $9.95.

BEST OF ABC'S WIDE WORLD OF SPORTS: THE '60S. The popular sports TV program began in 1960; this chronicles its first decade. (CBS, 1990) 67 min., $19.98.

BEST OF ABC'S WIDE WORLD OF SPORTS: THE '70S. The thrill of victory and the agony of defeat in its second decade. (CBS, 1990) 70 min., $19.98.

BEST OF ABC'S WIDE WORLD OF SPORTS: THE '80S. As sports became a bigger business, Wide World grew along with it. (CBS, 1990) 68 min., $19.98.

BILL ELLIOTT: RACING INTO HISTORY. Inside look at one of NASCAR's most successful and popular drivers. $9.95.

BIZARRE SPORTS AND INCREDIBLE FEATS. 30 min., $9.95.

CHRIS BERMAN UNBELIEVABLE SPORTS PLAYS. From the archives of ESPN, Boomer presents some of the most unusual plays from a variety of sports. (ESP) 30 min., $9.95.

ESCAPE TO SKI. From Maui to Malibu, Warren Miller's film features deep-powder helicopter skiing in western Canada, spring skiing in France, snowboards in Colorado, and pro racing in California. (WMH, 1989) 90 min., $39.95.

GET THE FEELING: POWER (*Sports Illustrated*). Experience the feeling of raw power through footage of football, boxing, etc. (HBO, 1988) 60 min., $9.99.

GET THE FEELING: SPEED (*Sports Illustrated*). Experience exciting sports footage on the feeling of speed. (HBO, 1987) 60 min., $9.99.

GET THE FEELING: Winning (*Sports Illustrated*). Experience exciting film footage on the feeling of winning. (HBO, 1988) 60 min., $9.99.

GOLD MEDAL CHAMPIONS. Highlights of American athletes at the Olympics. $14.95.

GREAT SPORTS MOMENTS OF THE '80S. Everything from the America's Cup to the Boston Marathon; star performances in boxing baseball, football, you name it. (CBS, 1990) 45 min., $19.98.

GREATEST MOMENTS IN CHICAGO SPORTS HISTORY. As selected by Chicago *Tribune* readers; includes 1906 World Series, Bears' Super Bowl win, Ernie Banks, Nellie Fox, more. (NFL, 1987) 55 min., $24.95.

GREATEST MOMENTS IN PHILADELPHIA SPORTS HISTORY. As selected by readers of the Philadelphia Daily News; includes Wilt Chamberlain, Chuck Bednarik, Dr. J., Bobby Clarke, more. (NFL) 45 min., $24.95.

GREATEST SPORTS LEGENDS: HALL OF FAME. Featuring 16 of sports all-time greatest performers; Ruth, Jim Brown, Palmer, etc. (RVI) 30 min., $9.98.

GREATEST SPORTS LEGENDS: MIRACLE MOMENTS OF SPORT. Some of the greatest accomplishments in sports history; Hank Aaron's 715th homer, Jerry West's miracle shot, Mark Spitz's seven gold medals, more. (RVI) 30 min., $9.98.

GREATEST SPORTS LEGENDS 10TH ANNIVERSARY SPECIAL: 101 SUPERSTARS. 101 superstars from baseball, basketball, football, etc. (RVI) 67 min., $9.98.

GREATEST SPORTS LEGENDS: 160 SUPERSTARS OF THE 20TH CENTURY. Hosted by Johnny Bench; whirlwind but fascinating tour. (RVI) 48 min., $9.98.

GREATEST SPORTS LEGENDS: RECORD BREAKERS, VOLUME I. See DiMaggio, Shoemaker, Owens, and others set the records that made them great. (RVI) 32 min., $9.98.

HARD ROAD TO GLORY. History of the black athlete in America; Joe Louis, Jesse Owens, Jackie Robinson, more. (WKV) 60 min., $19.95.

HEROES AND HEARTACHES: *A Treasury of Boston Sports Since 1975.* The Sox, the Celtics, the Marathon, more. (MLB, 1987) 60 min., $19.95.

HEROINES: *Early Women Sports Stars.* Women athletes from the 1920s through the '50s. 60 min.

HIGHLIGHTS OF THE 1988 SUMMER OLYMPICS (Seoul). 90 min., $29.95.

THE HISTORY OF THE INDIANAPOLIS 500. Video retrospective covering 75 years of racing at Indy. (JCI) $19.95.

ICE SKATING SHOWCASE: *Great Routines of the '80S.* Fourteen of the finest skating routines of the decade. 60 min., $19.95.

JESSE OWENS. Greatest Sports Legends. (RVI) 30 min., $9.98.

JEWELS OF THE TRIPLE CROWN. Story of the eleven Triple Crown winners; written and narrated by ABC's Jim McKay. (CBS) 60 min., $19.98.

LIVE AND DRIVE THE INDY 500. 30 years of rare and exciting footage, stories of famous drivers, etc. 60 min., $19.98.

MAGIC MEMORIES ON ICE. Three decades of outstanding skating. (CBS, 1989) 90 min., $24.95.

MAGICIANS OF SPORT. Featuring some of the cleverest athletes in the most magical moments. (HBO) 55 min., $14.99.

OFFICIAL 1988 CALGARY WINTER OLYMPICS VIDEO. (CBS) $24.95.

OLYMPIC BOXING. 1988 Seoul. 45 min., $19.95.

OLYMPIC GOLD MEDAL WINNERS: *The First 90 Years.* Highlights of the great gold medal winners in the first 90 years of the modern games. 40 min., $19.95.

OLYMPIC GYMNASTICS. 1988 Seoul. 45 min., $19.95.

OLYMPIC TRACK & FIELD (Men). 1988 Seoul. 45 min., $19.95.

OLYMPIC TRACK & FIELD (Women). 1988 Seoul. 45 min., $19.95.

OLYMPIC VOLLEYBALL. 1988 Seoul. 45 min., $19.95.

OLYMPIC WATER SPORTS. 1988 Seoul; Swimming & Diving. 45 min., $19.95.

RARE SPORTSFILMS (RSF) AUTO RACING: $29.95 each.

> 1930s-1964 "NEWS FROM INDY." Rare newsfilm footage of the 500, B&W, 45 min.
>
> 1949 INDY "BEHIND THE CHECKERED FLAG." Bill Holland winner. CO, 28 min.
>
> 1952 INDY "THE FABULOUS 500." Troy Ruttman becomes youngest winner. CO, 25 min.
>
> 1953 INDY "THE HOTTEST 500." Bill Vukovich wins his first 500. CO, 26 min.
>
> 1954 INDY "THE FANTASTIC 500." Vukovich wins second in a row. CO, 30 min.
>
> 1964 INDY "WAY OF A CHAMPION." The Granatelli STP Novi Story. B&W, 29 min.
>
> 1956 "RACING TRIPLEHEADER." '56 Indy, '56 Daytona, & '55 Southern 500. CO, 28 min.

General, Olympics and Other Sports 185

1951-59 "PIONEERS OF STOCK CAR RACING." Highlights of Grand National Races. B&W, 41 min.

1952 DAYTONA. Live driver interviews and the '52 classic run on the beach. B&W, 26 min.

1956 SOUTHERN 500 AT DARLINGTON, S.C. Curtis Turner wins the big one! CO, 29 min.

1958 DAYTONA. Goldsmith wins last race on the old beach course. B&W, 26 min.

1961 SOUTHERN 500 AT DARLINGTON, S.C. Nelson Stacy beats out Roberts/Panch. CO, 30 min.

1963 RIVERSIDE 500. First 500-mile stock car road race ever run. CO, 36 min.

1963 ATLANTA 500 AT ATLANTA INTERNATIONAL RACEWAY. Fred Lorenzen winner. CO, 28 min.

1963 SOUTHERN 500 AT DARLINGTON. Fireball Roberts' last superspeedway win. CO, 29 min.

1970 GRAND NATIONAL HIGHLIGHTS. Superbirds and Dodge Daytonas plus Bobby Isaac. CO, 28 min.

1971 SOUTHERN 500 AT DARLINGTON. Bobby Allison's first Southern 500 win. CO, 29 min.

RECORD BREAKERS OF SPORT. (RVI) 51 min., $14.99.

RICHARD PETTY. Greatest Sports Legends. (RVI) 30 min., $9.98.

ROCK CHALK, JAYHAWK. Athletic history of the University of Kansas. $19.95.

16 DAYS OF GLORY. A closeup look at the 1984 Summer Olympics. (PAR) 145 min., $24.95.

16 DAYS OF GLORY II. More highlights from the 1984 Summer Olympics. (PAR) 147 min., $24.95.

SKIING EXTREME: *Hot Music! Radical Runs! Big Air!* 22 min., $19.95.

SPORTS COLOSSUS: *Heroes of the '20S, '30S, and '40S.* Hosted by Bob Mathias. 109 min., $24.95.

SPORTS HEROES. Highlights of some of sports' greatest performers. (RVI) 30 min., $9.95.

SPORTS IQ TEST. Sports trivia for experts. 30 min., $9.95.

25 YEARS OF SPORTS: 1943-67. Taken from newsreels of sports' greatest moments. 60 min., $14.95.

TIME CAPSULE: LOS ANGELES OLYMPIC GAMES OF 1932. A recreation, employing actors, of the milieu of the Games. 60 min., $29.95.

TIME WAITS FOR SNOWMAN. A high-flying ski thriller., $19.95.

WARREN MILLER'S EXTREME SKIING. Cliff jumps, world class snowboarders, outrageous stunts. (WMH, 1990) 45 min., $19.95.

WARREN MILLER'S SKI COUNTRY. Twenty-one famous ski resorts, from New Zealand to the Alps. (WMH, 1987) 96 min., $19.95.

WARREN MILLER'S SKI TIME: *A Ski Vacation in a Box* Ski the world's greatest mountains. (WMH, 1988) 107 min., $19.95.

WARREN MILLER'S SNOWONDER. The world's most exciting ski terrain. (WMH) 98 min., $19.95.

WARREN MILLER'S STEEP AND DEEP. (WMH) 90 min., $19.95.

WARREN MILLER'S STEEPS, LEAPS AND POWDER. The world's greatest extreme skiers leap from trams, glide over crevasses, and ski under the best and worst conditions. (WMH, 1989) 50 min., $24.95.

WARREN MILLER'S WHITE WINTER HEAT. Above 10,000 feet, beyond the realm of the ordinary, skiing in Switzerland, British Columbia, Montana, and Argentina. (WMH, 1988) 87 min., $59.95.

Instructionals (by sport)

Archery

ARCHERY SERIES: *Invitation to Archery.* Archery fundamentals, intermediate techniques, etc. 45 min., $39.00.

Billiards

AN AFTERNOON WITH GERRY WATSON: *Fundamental Techniques and Professional Secrets* (Billiards). Hosted by Canada's premier billiards entertainer. 30 min., $9.95.

BYRNE'S STANDARD VIDEO OF POOL AND BILLIARDS. A basic, step-by-step guide from basic to advanced techniques. 60 min., $29.95.

BYRNE'S STANDARD VIDEO OF POOL AND BILLIARDS, VOL. II. Continuing instruction., $29.95.

HOW TO PLAY POOL WITH MINNESOTA FATS. Instructional. 60 min., $19.98.

Bowling

BOWL TO WIN WITH EARL ANTHONY. Instructional. $19.95.

BOWLING SERIES: *New Approach to a Great Old Game.* Fundamentals of the arm swing, footwork, strikes, and spares. 40 min., $29.95.

BOWLING WITH NELSON BURTON, JR. Inside tips for new and experienced bowlers. (WKV) 60 min., $19.95.

EARL ANTHONY: GOING FOR 300. Not batting average, bowling score. $19.95.

LET'S BOWL WITH DICK WEBER. Instructional. 40 min., $16.95.

SCORE MORE BOWLING WITH NELSON BURTON, JR.. Instructional. (1988) 60 min., $24.98.

STRIKE: *The Guide to Consistent Bowling with Joe Berardi* 40 min., $24.95.

TEACHING KIDS BOWLING. Gordon Vadakin teaches the basics. (ESP) 40 min., $29.95.

Field Hockey

HOCKEY: THE BASIC SKILLS. Field hockey instructional. 45 min., $29.95.

Gymnastics

GYMNASTICS FUN WITH BELA KAROLI. Learn from the coach of Olympic Gold Medalists Mary Lou Retton and Nadia Comaneci. (VES, 1988) 60 min., $29.98.

Horse Racing

HORSES TALK: *The Paddock and Post Parade.* Companion piece to the book of the same name; explains what to look for in a horse before the race. $39.95.

Ice Skating

HOW TO ICE SKATE. Instructional video for beginners through advanced skaters, as taught by Tai Babilonia and Randy Gardner. (FRI, 1986) 60 min., $24.95.

Lacrosse

BRINE LACROSSE CLINIC. Instructional by coaches Don Zimmerman of Johns Hopkins and Richie Moran of Cornell. 29 min., $29.95.

Running / Track and Field

CHAMPIONSHIP VIDEO SERIES: SPRINTS AND RELAYS. 48 min., $49.95.

DO IT BETTER: RUNNING. Asics Sports Video Collection; with marathoner Ingrid Kristiansen and miler Steve Scott. 30 min., $24.95.

RUNNING GREAT WITH GRETE WAITZ. Tips from the great marathoner. 60 min., $14.95.

SEBASTIAN COE: BORN TO RUN. Program for coaches and athletes based on Coe's own mental and physical preparation. 52 min., $39.95.

SPEED AND EXPLOSION. Speed improvement for athletes. 53 min., $39.95.

SPRINTING WITH CARL LEWIS AND COACH TOM TELLEZ. A step-by-step instruction on sprinting. 58 min., $39.95.

TEACHING KIDS SPEED FOR ALL SPORTS WITH CARL LEWIS. 45 min., $19.95.

TRACK & FIELD: COACHING BY THE EXPERT OLYMPIC TEAM COACHES SERIES. 40 min., $39.95 each.

   I: RUNNING EVENTS. With LeRoy Walker, Larry Ellis, Stan Huntsman, and Terry Crawford.

   II: JUMPING EVENTS. With Tom Tellez.

   III: THROWING EVENTS. With Tom Pagoni.

WORLD CLASS TRACK AND FIELD: HIGH JUMP. With Hollis Conway and Coach Dick Booth. $39.95.

Skiing

BILLY KIDD SKI RACING: *The Fast Way to Improve Your Skiing.* 27 min., $14.95.

BODY PREP: *The Ultimate Ski Fitness Video.* Women pro racers present a year-round ski fitness program. (WMH, 1989) 101 min., $29.95.

PETER BOGNER'S SKIING TECHNIQUES. Covers free skiing and recreational racing for the intermediate to advanced skier. (IVE, 1987) 45 min., $19.95.

BREAKTHROUGH BASICS OF DOWNHILL SKIING WITH HANK KASHIWA. 45 min., $29.95.

CROSS-COUNTRY SKIING BASICS. Bill Koch travels Oregon's Mt. Bachelor for a primer on fundamentals. (ESP, 1988) 48 min., $29.95.

DISTINCTIVE SKIING. How to beautify your skiing. 82 min., $39.95.

DOWNHILL SKIING BASICS. Ski instructor Hank Kashiwa teaches beginners how to start off right. (1988) 45 min., $29.95.

THE DOWNHILL SKIING PRIMER. Ski pro Gary Skoog gives lessons you can follow right in your own living room. (MR, 1987) 30 min., $9.95.

SKATING AWAY: CROSS-COUNTRY SKIING. Olympic medalist Bill Koch ranges the mountains of New Zealand. (ESP, 1988) 48 min., $29.95.

SKI THE MAHRE WAY. Olympic skiing greats Steve and Phil Mahre provide professional guidelines for downhill skiers from beginners to experts. (1988) 55 min., $24.95.

SKIING WITH STYLE: *Mastering the Mountain.* 60 min., $39.95.

TEACHING KIDS SKIING. Hank Kashiwa prepares children for their first outing on the slopes. (ESP, 1988) 45 min., $29.95.

WARREN MILLER'S LEARN TO SKI BETTER. Tips on all aspects of skiing. (WMH) 85 min., $19.98.

Softball

AMATEUR SOFTBALL ASSOCIATION VIP SOFTBALL SERIES

ADVANCED FASTBALL PITCHING. Shows grips and releases for various pitches, problem-solving tips, and practice hints. 75 min., $29.95.

DEFENSIVE FUNDAMENTALS AND DRILLS. Keys to successful fielding, both infield and outfield. 60 min., $29.95.

SLOW PITCH TEAM DEFENSE, STRATEGY AND SLIDING. Defensive coverage. 60 min., $29.95.

FAST PITCH TEAM DEFENSE, STRATEGY AND SLIDING. Defensive coverage. 30 min., $29.95.

CHAMPIONSHIP SOFTBALL HITTING SYSTEM. With Bob Campbell. 30 min., $9.99.

DO IT BETTER: SLO PITCH SOFTBALL. Asics Sports Video Collection; Hall of Famers show techniques for hitting and pitching. 30 min., $24.95.

PITCHING SLOW PITCH SOFTBALL. Pitches, releases, and defense. 45 min., $9.99.

POWER HITTING IN SOFTBALL. Tips from the best hitters in the sport. 25 min., $9.95.

SLO PITCH SOFTBALL: REFLEX HITTING SYSTEM WITH RAY DeMARINI. Covers stance, swing, bat speed, etc. 60 min., $29.95.

SOFTBALL: PUTTING IT TOGETHER. U. of Minnesota coach Linda Wells covers all aspects of women's softball. 69 min., $45.00.

SOFTBALL SERIES: BASIC SKILLS IN SOFTBALL. Better hitting, baserunning, pitching. 33 min., $29.95.

STRATEGY OF PITCHING SLO PITCH SOFTBALL. 45 min., $29.95.

Swimming and Diving

DIVING MY WAY. With Ron O'Brien, U.S. Olympic Diving Coach. 98 min., $39.95.

DON GAMBRIL'S CLASSIC SERIES. Led by Alabama and former Olympic team coach. 60 min., $39.95 each.

  MEN'S SWIMMING

  WOMEN'S SWIMMING

DONNA DeVARONA'S NO IMPACT WORKOUT: SWIM YOUR WAY TO FITNESS. 53 min., $19.95.

SWIM LESSONS FOR KIDS: A SIMPLE, PROVEN METHOD FOR PARENTS TO TEACH THEIR CHILDREN OVER THREE TO SWIM. Includes illustrated booklet. 40 min., $29.95.

SWIM SMARTER, SWIM FASTER. Stroke technique for swimmers from 8 to 80. $29.95.

SWIMMING: EXCELLENCE IN SWIMMING STROKE TECHNIQUE. Freestyle, backstroke, breaststroke, and butterfly, starts and turns with national and Olympic coach Mark Schubert. 90 min., $39.95.

TEACHING KIDS SWIMMING WITH JOHN NABER. Teaching children the basics of swimming and water safety. 42 min., $29.95.

WATER IS FRIENDLY: *First Step in Learning to Swim.* 34 min., $49.95.

Triathlon

TRIATHLON TRAINING AND RACING WITH DAVE SCOTT. (FRI) 87 min., $12.95.

Volleyball

COACH TO WIN: *Coaching Middle School and Elementary Volleyball.* Hosted by Ball State Coach Don Shondell. 60 min., $39.95.

DO IT BETTER: BEACH VOLLEYBALL. Asics Sports Video Collection; with UCSB women's volleyball coach Kathy Gregory. 45 min., $24.95.

DO IT BETTER: VOLLEYBALL. Asics Sports Video Collection; setting, serving, individual defense, blocking, spiking, etc.; with U. of Hawaii coach Don Shoji. 45 min., $24.95.

Wrestling, Amateur

DO IT BETTER: WRESTLING. Asics Sports Video Collection; with Arizona St. coach Bob Douglas. 30 min., $24.95.

KIDS WRESTLING SERIES. Tips on the how-to's of kid wrestling; includes easy to follow instructional manual. 45 min., $29.95 each.

   PART 1: ORGANIZING A KIDS' WRESTLING CLUB.
   PART 2: KEYS TO FITNESS, NUTRITION, AND SAFETY.
   PART 3: BASIC SKILLS AND BETTER TECHNIQUES.

Bloopers and Humor

ALL NEW NOT-SO-GREAT MOMENTS IN SPORTS. Sequel to the best-selling assortment of sports bloopers. (HBO, 1988) 45 min., $14.98.

ALL NEW BOB UECKER'S WACKY WORLD OF SPORTS. Sequel to his original "Wacky World" video; Combination of bloopers and Uecker's off-the-wall humor. (RVI, 1986) 30 min., $9.95.

AMAZING BIF BAM BOOM ANYTHING GOES SPORTS BLOOPERS. Roy Firestone hosts a collection of bloopers from a variety of sports. (ESP, 1989) 45 min., $9.95.

BEST OF BOB UECKER'S WACKY WORLD OF SPORTS. The catcher-turned-comedian hosts a mix of sports bloopers and his own humor. 30 min., $9.98.

THE BEST OF COMEDY, VOLUME 1. Warren Miller's collection of unfortunate ski spills, unplanned accidents, and blunders. (WMH, 50 min., $24.95.

BIZARRE SPORTS AND INCREDIBLE FEATS. 30 min., $9.95.

DORF AND THE FIRST GAMES OF MT. OLYMPUS. Tim Conway's Dorf character branches out from golf and recreates the first Olympic games—well, sort of. (J2) $29.95.

GOLDEN GOOFY CLASSICS. Vintage footage of incredible acts of courage and stupidity. 30 min., $14.95.

GREATEST SPORTS FOLLIES. Lowlights from football, basketball, and many other pastimes. (CBS, 1980) 45 min., $14.98.

THE NOT-SO-GREAT MOMENTS IN SPORTS. Best-selling video full of moments many leading sports personalities would love to forget. (HBO, 1987) 54 min., $9.95.

THE NOT-SO-GREAT MOMENTS IN SPORTS, TAKE 3. The 1990 edition of the classic. (HBO) 46 min., $14.98.

PRO SPORTS BLOOPERS. 30 min., $9.95.

SPORTS BLOOPERMANIA. Featuring Morgana the "Kissing Bandit" and assorted bloopers from the world of sports. 30 min., $14.95.

SPORTS BLOOPERS AWARDS WITH CHRIS BERMAN. The best bloopers from ESPN's library. (ESP) 35 min., $9.95.

SUPER SPORTS FOLLIES. Includes rodeos, boxing, lacrosse, practical jokes, etc. 30 min., $9.95.

TIME OUT FOR HILARIOUS SPORTS BLOOPERS. Narrated by Roy Firestone. 30 min., $14.95.

WARREN MILLER'S SPORTS BLOOPERS. Foulups and pratfalls by America's weekend athletes. (WMH) $24.95.

WORLD WIDE SPORTS BLOOPERS. 30 min., $9.98.

MOVIES

The mere appearance of a shoulder pad or a basketball does not make a sports movie. The climax of Blake Edwards' *Experiment in Terror* takes place on the pitcher's mound at Candlestick Park, but the film is about a terroristic bank robber, not baseball.

Nor is it enough to have a character who is purportedly an athlete. Lee Marvin is an ex-baseball player in *Ship of Fools* and Paul Newman an ex-footballer in *Cat on a Hot Tin Roof*, but neither film has any sports pretensions.

On the other hand, few movies that I would classify as the "sports" type are actually *about* the sport they depict. They are almost always about love, or coming back from a disability, or male bonding, or else they are gimmicky comedies or splashy musicals. The baseball played in *Damn Yankees* could hardly serve as an instructional.

To make our list, the characters should spend a significant amount of time playing, talking about, or watching a sport, no matter what the movie is really about. To be honest, I've sometimes been a little generous with that "significant time" factor so I could add in a film I thought you might like to know about.

There are a number of useful books available that list just about every movie available in the world, but you have to read them from cover to cover to find all the sports films. Here, I've listed sports films alphabetically according to sport. I coded them as to quality (# = awful, * = poor, ** = average, *** = good, **** = excellent) as *movies*, not as to how realistically they depict certain sports.

Below the titles are the year released; the motion picture rating (PG, etc.); whether it's in color (C), black and white (B&W), or colorized (CZ); the running time in minutes; whether it was made for TV (TVM); and if we could verify that it is available on video (+++). We listed some movies that have not been released as videos at this writing because they may be someday (or you might want to set your VCR for the Late Show).

After the director (Dir.) and cast, you'll find a capsule comment that may entice you into watching or warn you off.

Baseball Movies

Ever since Ty Cobb was talked into making *Somewhere in Georgia* in 1916, Hollywood has been trying to make a good baseball movie. In the good ones of a few decades ago, they wrapped the baseball in a heartwarming story (*Pride of the Yankees*, *The Stratton Story*), a screwball comedy (*It Happens Every Spring*, *Rhubarb*), or a musical (*Damn Yankees*, *Take Me Out to the Ball Game*. Baseball sat on the bench while sentiment or chuckles or songs took the field.

Only in the 1980s did some people who actually liked baseball and had faith in it get to make films. The pleasant result has been such superior efforts as *Eight Men Out*, *The Natural*, *Field of Dreams*, and, best of all, *Bull Durham*.

ALIBI IKE **½
(1935) B&W 73 min.
Dir.: Ray Enright. Cast: Joe E. Brown, Olivia de Havilland, William Frawley, Ruth Donnelly, Roscoe Karns.

Fine retelling of Ring Lardner's story of a bragging, tall-tale-telling pitcher. One of Joe E.'s best, and the young de Havilland is charming as his girl.

ANGELS IN THE OUTFIELD **
(1951) B&W 102 min.
Dir.: Clarence Brown. *Cast*: Paul Douglas, Janet Leigh, Keenan Wynn, Donna Corcoran.

Whimsy fairly drips from this tail of a waif who's the only one who can see angels helping the bungling Pittsburgh Pirates win games. The movie's better than the real Pirates of the time, and Douglas is fun as a foulmouthed (for 1951) manager, but take your insulin before watching this "Miracle at Forbes Field."

AUNT MARY ***
(1979) C 100 min. TVM +++
Dir.: Peter Werner. *Cast*: Jean Stapleton, Martin Balsam, Harold Gould, Dolph Sweet, Robbie Rist, Anthony Cafiso, K.C. Martel.

Mary Dobkin was a physically handicapped Baltimore woman who became a legendary coach in sandlot baseball. A good cast does her story justice in this inspirational made-for-TV film.

THE BABE RUTH STORY ½
(1948) B&W 106 min. +++
Dir.: Roy Del Ruth. *Cast*: William Bendix, Claire Trevor, Charles Bickford, Sam Levene, William Frawley.

Ludicrous biography of the Sultan of Swat should have been swatted in the planning stage. Some unintended laughs make it almost—but not quite—worth watching.

THE BAD NEWS BEARS ***
(1976) [PG] C 91 min. +++
Dir.: Michael Ritchie. *Cast*: Walter Matthau, Tatum O'Neal, Vic Morrow, Joyce Van Patten, Jackie Earle Haley, Alfred W. Lutter. Excellent comedy about an inept and foulmouthed Little League team that finds a winning combination in a beer-guzzling manager (Matthau) and a girl pitcher (O'Neal). Great fun.

THE BAD NEWS BEARS GO TO JAPAN ½
(1978) [PG] C 91 min. +++
Dir.: John Berry. *Cast*: Tony Curtis, Jackie Earle Haley, Tomisaburo Wakayama, George Wyner, Lonnie Chapman.

The kind of movie that gives sequels a bad name. Curtis is a hustler trying to make a buck on the Bears and—oh, to hell with it. This was the last of a series that went downhill like a bobsled.

THE BAD NEWS BEARS IN BREAKING TRAINING *
(1977) [PG] C 100 min. +++
Dir.: Michael Pressman. *Cast*: William Devane, Jackie Earle Haley,

Jimmy Baio, Clifton James, Chris Barnes.

Poor sequel to the classic Matthau comedy, has only a few laughs. Haley, star of Little League team, goes to Astrodome and talks estranged father (Devane) into coaching the brats. Yuk!

BANG THE DRUM SLOWLY ***
(1973) [PG] C 97 min. +++
Dir.: John Hancock. *Cast*: Michael Moriarty, Robert De Niro; Vincent Gardenia, Phil Foster, Ann Wedgeworth, Patrick McVey, Heather MacRae, Selma Diamond, Barbara Babcock, Tom Ligon, Nicholas Surovy, Danny Aiello.

Moriarty is a star pitcher; De Niro is a sub catcher. They come together in a bond of friendship when the catcher learns he is dying. Touching screen play by Mark Harris based on his 1956 novel. Both leads are terrific.

THE BINGO LONG TRAVELING ALL-STARS AND MOTOR KINGS ***
(1976) [PG] C 110 min. +++
Dir.: John Badham. *Cast*: Billy Dee Williams, James Earl Jones, Richard Pryor, Ted Ross, DeWayne Jesse, Stan Shaw.

Williams tries to buck Negro Baseball League owners in 1939 by starting his own team. A bright, original comedy well worth your time. Some ex-Negro Leaguers were reportedly critical that so much emphasis is placed on the clowning that black teams often had to do to draw crowds in those days.

BULL DURHAM ****
(1988) [R] C 108 min. +++
Dir.: Ron Shelton. *Cast*: Kevin Costner, Susan Sarandon, Tim Robbins, Trey Wilson, Robert Wuhl, Johnny Robertson, Max Patkin.
Story of a career minor league catcher (Costner), baseball groupie (Sarandon), and talented but screwball pitcher (Robbins) is the most honest (and, to my mind, best) baseball movie ever. Nearly every moment rings real: Costner talking himself through an at bat, Wuhl's dugout chatter, an "unexpected" open date, the conference on the mound, and on and on. Funny, true-to-life, and, at times, world-class sexy.

THE COMEBACK KID *
(1980) C 97 min. +++
Dir.: Peter Levin. *Cast*: John Ritter, Doug McKeon, Susan Dey, Jeremy Licht, James Gregory.

Ritter is a down-and-out ballplayer who ends up coaching some street kids in this mail-it-in made-for-TV. Don't come back!

DAMN YANKEES ***
(1958) C 110 min. +++
Dir.: George Abbott, Stanley Donen. *Cast*: Tab Hunter, Gwen Verdon, Ray Walston, Russ Brown, Jimmie Komack, Jean Stapleton, Nathaniel Frey.

Musical variation on the *Faust* story: an aging Washington fan sells his soul to the devil and becomes the star (Hunter) to lead the Senators to victory over the Yankees—maybe. No real baseball, but Verdon and Walston (from the original Broadway hit) are terrific, the music ("You Gotta Have Heart," "Whatever Lola Wants," etc.) is great, and the whole thing makes for a good time.

EIGHT MEN OUT ***½
(1988) [PG] C 119 min. +++
Dir.: John Sayles. *Cast*: John Cusack, Clifton James, Michael Lerner, Christopher Lloyd, John Mahoney, Charlie Sheen, David Strathairn, D.B. Sweeney, Don Harvey, Michael Rooker, Perry Lang, James Read, Bill Irwin, Keven Tighe, Studs Terkel, John Anderson, Maggie Renzi.

This story of the 1919 Black Sox throwing the World Series is faithful to the period and to Eliot Asinof's book. The baseball is believable and the performances, especially David Strathairn as Eddie Cicotte, are all on the mark. Yet, the result is curiously uninvolving. Where's the center? So many stories are part of this tragedy that none really grabs the viewer. A noble effort that might have been the best ever had it told a little less and concentrated on a couple of major characters.

ELMER THE GREAT **½
(1933) B&W 74 min.
Dir.: Mervyn LeRoy. *Cast*: Joe E. Brown, Patricia Ellis, Claire Dodd, Sterling Holloway, Jessie Ralph.

Film version of Ring Lardner story in which naive Brown becomes a baseball star and is taken advantage of by crooks. Dated but amusing. Brown was an athlete who often worked out with real professional teams.

FEAR STRIKES OUT ***
(1957) B&W 100 min. +++
Dir.: Robert Mulligan. *Cast*: Anthony Perkins, Karl Malden, Norma Moore, Adam Williams, Perry Wilson.

Based on Jim Piersall's own account of his mental breakdown and recovery as a major league rookie. Perkins is intense though not the most convincing athlete in the world; Malden is great as his domineering father.

FIELD OF DREAMS ***
(1989) [PG] C 106 +++
Dir.: Phil Alden Robinson. Kevin Costner, Amy Madigan, Gaby Hoffman, Ray Liotta, Timothy Busfield, James Earl Jones, Burt Lancaster, Frank Whaley, Dwier Brown.

Popular fantasy about the Iowa farmer who constructs a stadium in his cornfield because a "voice" tells him to—"If you build it, they will come." More baseball is talked than played ("If you film it, they will gab"), but this story of redemption pushes most of the right buttons for maximum impact. Based on W.P. Kinsella's book *Shoeless Joe*.

THE GREAT AMERICAN PASTIME **
(1956) B&W 89 min.
Dir.: Herman Hoffman. *Cast*: Tom Ewell, Anne Francis, Ann Miller, Dean Jones, Ruby Dee.

Ewell is very good in this quiet little comedy about the effect on his family life caused by his coaching a Little League team. Not as funny or well done as "Bad News Bears," but serviceable.

IT HAPPENS EVERY SPRING ***
(1949) B&W 87 min.
Dir.: Lloyd Bacon. *Cast*: Ray Milland, Jean Peters, Paul Douglas, Ed Begley.

Milland is a college chemistry professor who accidentally invents a mixture that avoids wood—like bats, when it's rubbed on a baseball—that lets him becomes an overnight major league pitching star. Highly enjoyable comedy, with Douglas nearly stealing the show as Milland's gruff, heart-of-gold catcher.

IT'S GOOD TO BE ALIVE **½
(1974) C 100 TVM
Dir.: Michael Landon. *Cast*: Paul Winfield, Lou Gossett, Ruby Dee, Ramon Bieri, Joe DeSantis, Ty Henderson, Lloyd Gow.

Story begins with the crippling accident that ended Roy Campanella's career and follows his inspiring recovery as a person. Winfield, Gossett, and Dee are excellent.

THE JACKIE ROBINSON STORY ***
(1950) B&W 76 min. +++
Dir.: Alfred E. Green. *Cast*: Jackie Robinson, Ruby Dee, Minor Watson, Louise Beavers, Richard Lane, Harry Shannon, Ben Lessy, Joel Fluellen.

Every baseball fan should see this fine piece of social history. Robinson's heroic struggle in breaking baseball's color bar after World War II is a lesson every generation should learn. But this isn't just a dose of medicine; the movie works as movie and Robinson is quite good at playing himself, no easy task.

THE KID FROM CLEVELAND #
(1949) B&W 89 min.
Dir.: Herbert Kline. *Cast*: George Brent, Lynn Bari, Rusty Tamblyn, Tommy Cook, Ann Doran.

Sportscaster Brent attempts to salvage a troubled youth with the help of the Cleveland Indians 1948 Championship team. Then-owner Bill Veeck threatened to seek and burn every print. That alone should have gained him Hall of Fame entrance.

THE KID FROM LEFT FIELD **
(1953) B&W 80 min.
Dir.: Harmon Jones. *Cast*: Dan Dailey, Anne Bancroft, Billy Chapin, Lloyd Bridges, Ray Collins, Richard Egan.

Small comedy with small laughs about ex-major leaguer (Dailey), reduced to being a ballpark vendor, who passes on tips through his son to change a losing team to a winner.

THE KID FROM LEFT FIELD *½
(1979) C 100 min. TVM +++
Dir.: Adell Aldrich. *Cast*: Gary Coleman, Robert Guillaume, Tab Hunter, Tricia O'Neil, Gary Collins, Ed McMahon.

Remake of the 1953 less-than classic. Why?

KILL THE UMPIRE *½
(1950) B&W 78 min.
Dir.: Lloyd Bacon. *Cast*: William Bendix, Una Merkel, Ray Collins, Gloria Henry, William Frawley, Tom D'Andrea.

Bendix is better as a fan who becomes a hated umpire than he was as Babe Ruth. (He could hardly help but be.) Some fair slapstick is the only reason to watch.

LONG GONE **½
(1987) C 110 min. TVM +++
Dir.: Martin Davidson. *Cast*: William L. Petersen, Virginia Madsen, Dermot Mulroney, Larry Riley, Katy Boyer, Henry Gibson, Teller.

Entertaining comedy about life in the minor leagues in 1950s holds up well until the too-pat ending. Nevertheless, there's a lot of believable baseball and Petersen scores as the team's player-manager.

A LOVE AFFAIR: THE ELEANOR AND LOU GEHRIG STORY **½
(1978) C 96 min. TVM +++
Dir.: Fielder Cook. *Cast*: Blythe Danner, Edward Herrmann, Patricia Neal, Jane Wyatt, Gerald S. O'Loughlin, Ramon Bieri, Georgia Engel, David Ogden Stiers, Lainie Kazan.

"Pride of the Yankees" told from Mrs. Gehrig's point of view. Danner and Herrmann are excellent in the title roles.

MAJOR LEAGUE **½
(1989) [R] C 107 min. +++
Dir.: David S. Ward. *Cast*: Tom Berenger, Charlie Sheen, Corbin Benson, Margaret Whitton, James Gammon, Rene Russo, Wesley Snipes, Dennis Haysbert, Charles Cyphers, Bob Uecker.

The Cleveland Indians' owner wants to make the team so bad that no one will come (Hey! Is this a documentary?) so she can move the franchise. Naturally, a group of assorted oddballs and misfits make the Indians a pennant contender. (Or is it a fantasy?) Comedy tries to ride *Bull Durham*'s coat tails, but lacks the reality or wit.

MAKE MINE MUSIC (CASEY AT THE BAT SEGMENT) **½
(1946) C 74 min.
Dir: Joe Grant (Production Supervisor). Voice of Jerry Colonna.

Animated Walt Disney opus and, overall, not one of his best. In ten segments, ranging from very good to overly pretentious, but the "Casey at the Bat," with Colonna's narration, is one of the better parts.

THE NATURAL ***½
(1984) [PG] C 134 min. +++
Dir.: Barry Levinson. *Cast*: Robert Redford, Robert Duvall, Glen Close, Kim Basinger, Wilford Brimley, Richard Farnsworth, Barbara Hershey, Robert Prosky, Darren McGavin, Joe Don Baker.

This free adaptation of Bernard Malamud's novel isn't everyone's favorite. The story of a naturally gifted athlete becoming a belated major league star is long, sometimes murky, and fraught with symbolism. But it's also fascinating, thought-provoking, beautifully filmed, and capped by a rousing finale. Redford actually looks like a decent player, too.

ONE IN A MILLION: THE RON LeFLORE STORY **½
(1978) C 100 min. TVM
Dir.: William A. Graham. *Cast*: LeVar Burton, Madge Sinclair, Paul Benjamin, James Luisi, Billy Martin, Zakes Mokae.

Good job of depicting LeFlore's rise from armed robbery conviction to major league ballplayer. Well acted by Burton and the rest.

THE PRIDE OF ST. LOUIS **
(1952) B&W 93 min. +++
Dir.: Harmon Jones. *Cast*: Dan Dailey, Joanne Dru, Richard Crenna, Hugh Sanders.

Idealized biography of Dizzy Dean (Dailey), with enough real Diz stories to be sporadically amusing. Answer to a trivia question: Richard Crenna played his brother Paul.

THE PRIDE OF THE YANKEES ***
(1942) B&W 127 min. +++
Dir.: Sam Wood. *Cast*: Gary Cooper, Teresa Wright, Babe Ruth, Walter Brennan, Dan Duryea, Ludwig Stossel, Addison Richards, Hardie Albright.

Perhaps not as good as its reputation, but still very much worth seeing. Cooper is fine as doomed Yankees great Lou Gehrig; Wright is ideal as his wife. A bit overlong, but the memorable final sequence when Gehrig tells a packed Yankee Stadium that he's "the luckiest man on the face of the earth" could wring a tear from a granite statue.

RHUBARB **½
(1951) B&W 95 min.
Dir.: Arthur Lubin. *Cast*: Ray Milland, Jan Sterling, Gene Lockhart, William Frawley, Elsie Holmes, Leonard Nimoy.

Lightweight but entertaining comedy about a cat that inherits a baseball team. Based on an H. Allen Smith story.

SAFE AT HOME! #
(1962) B&W 83 min. +++
Dir.: Walter Doniger. *Cast*: Mickey Mantle, Roger Maris, William Frawley, Patricia Barry, Don Collier, Bryan Russell.

Only for rabid Yankee fans and masochists. A kid lies to his teammates that he knows Mantle and Maris; they save his psyche by showing up at his Little League banquet. The best acting is by the rubber chicken.

THE SLUGGER'S WIFE *
(1985) [PG] C 105 min. +++
Dir.: Hal Ashby. *Cast*: Michael O'Keefe, Rebecca DeMornay, Martin Ritt, Randy Quaid, Cleavant Derricks, Lisa Langlois, Louden Wainwright III.

Remarkably poor original screenplay by Neil Simon about a boorish ballplayer and the singer who loves him is given the production it deserves. Odious.

STEALING HOME *
(1988) [PG] C 98 min. +++
Dir.: Steven Kampmann, Will Aldis. *Cast*: Mark Harmon, Jodie Foster, William McNamara, Blair Brown, Harold Ramis, Jonathan Silverman, Richard Jenkins, John Shea, Ted Ross, Thatcher Goodwin, Yvette Croskey.

Thirty-something ballplayer (Harmon) is lured back into game he gave up fourteen years ago when he is gifted with the ashes of his suicided baby-sitter (Foster). Huh? Although some critics found merit here (and some baseball fans liked the St. Louis Browns), considering the talent wasted on this stupefying clunker, it's our nominee as the worst baseball-related movie ever made.

THE STRATTON STORY ***
(1949) B&W or CZ 106 min. +++
Dir.: Sam Wood. *Cast*: James Stewart, June Allyson, Frank Morgan, Agnes Moorehead, Bill Williams.

Sam Wood, who directed "Pride of the Yankees," has another heart-tugging winner in true story of pitcher Monte Stratton, who made a comeback in the Texas League after losing his leg in a hunting accident. Stewart and Allyson make an appealing couple, and watch for Yankee great Bill Dickey in a cameo.

TAKE ME OUT TO THE BALL GAME ***
(1949) C 93 min. +++
Dir.: Busby Berkeley. *Cast*: Frank Sinatra, Esther Williams, Gene Kelly, Betty Garrett, Edward Arnold, Jules Munshin, Richard Lane, Tom Dugan.

Tinker-to-Evers-to-Chance as a musical comedy. Turn-of-the-century baseball was never *really* like this, but who cares? Good songs, good dancing, some good laughs, and Williams in a wet bathing suit.

TIGER TOWN **½
(1983) [G] C 95 min. TVM +++
Dir.: Alan Shapiro. *Cast*: Roy Scheider, Justin Henry, Ron McLarty, Bethany Carpenter, Noah Moazezi, Mary Wilson.

Heartwarming (what else?) made-for-cable movie about young Tiger fan and the over-the-hill ballplayer he "wishes" to a final fine season. Okay, but no pennant-winner.

A WINNER NEVER QUITS **½
(1986) C 100 min. TVM +++
Dir.: Mel Damski. *Cast*: Keith Carradine, Mare Winningham, G.W. Bailey, Dennis Weaver, Fionnula Flanagan, Huckleberry Fox, Dana Delany, Charles Hallahan.

Pete Gray, who lost an arm in a childhood accident, was a minor league sensation and had one wartime season in the majors. Carradine is excellent in his depiction of Gray as more than a sideshow attraction.

THE WINNING TEAM **½
(1952) B&W 98 min.
Dir.: Lewis Seller. *Cast*: Ronald Reagan, Doris Day, Frank Lovejoy,

Eve Miller, James Millican, Russ Tamblyn.

Before he played the President he played Grover Cleveland Alexander! Alex was a great pitcher, but in real life not a very interesting person. This sporadically entertaining Hollywood-formula bio-pic had Mrs. Alex as adviser. Guess who comes off as the real star of the "team."

Basketball Movies

Considering all those stars who get themselves seen in the expensive seats at Lakers games, Hollywood has been sparing in its treatments of basketball. Perhaps the proportions of typical players don't lend themselves to the wide screen.

Of course, considering most of what they *did* make, we're better off. Well, *Hoosiers* is worth your time.

AMAZING GRACE AND CHUCK *
(1987) [PG] C 115 +++
Dir.: Mike Newell. *Cast*: Jamie Lee Curtis, Alex English, Gregory Peck, William L. Petersen, Dennis Lipscomb, Lee Richardson.

Well-intentioned bomb. A 12-year-old gives up sports to protest nuclear arms, and soon, boys and girls, all the high-priced athletes in the world—the ones who want to renegotiate their contracts every time they have a good game—join him in his sitdown. Amazing they made this movie-to-upchuck-by!

COACH
(1978) C 100 min. +++
Dir.: Bud Townsend. *Cast*: Cathy Lee Crosby, Michael Biehn, Keenan Wynn.

Bad exploitation film. Crosby is Olympics Gold Medal winner hired to coach sad-sack boys high school basketball. Air ball.

THE FISH THAT SAVED PITTSBURGH *
(1979) [PG] C 102 min. +++
Dir.: Gilbert Moses. *Cast*: Julius Erving, James Bond III, Stockard Channing, Jonathan Winters, Meadowlark Lemon, Flip Wilson,

Kareem Abdul-Jabbar.

Only a new script could save this movie! Bad team uses disco and astrology to win. Dr. J earns acting honors which is to his credit and everyone else's shame.

GO, MAN, GO **½
(1954) B&W 82 min.
Dir.: James Wong Howe. *Cast*: Dane Clark, Sidney Poitier, Pat Breslin.

Fictionalized story of the origins of the Harlem Globetrotters is worth **** when the team is in action and * for the rest. Our rating splits the difference.

HOOSIERS ****
(1986) [PG] C 114 min. +++
Dir.: David Anspaugh. *Cast*: Gene Hackman, Barbara Hershey, Dennis Hopper, Sheb Wooley, Fern Parsons.

Hackman is the new coach with a mysterious past who must redeem himself by leading his team to the 1951 state high school basketball championship. Hopper steals the show as an alcoholic hoops nut. Old-fashioned in the best sense. Hooray!

ONE ON ONE **½
(1977) [PG] C 98 min. +++
Dir.: Lamont Johnson. *Cast*: Robby Benson, Annette O'Toole, G.D. Spradlin, Gail Strickland, Melanie Griffith.

A basketball version of the nice-boy-adrift-in-corrupt-college sports. You've seen it all before, but Benson (who co-wrote script) does well as the kid.

THAT CHAMPIONSHIP SEASON **
(1982) [R] C 110 min. +++
Dir.: Jason Miller. *Cast*: Robert Mitchum, Bruce Dern, Stacy Keach, Martin Sheen, Paul Sorvino, Arthur Franz.

Broken dreams abound when a once-championship basketball team gets together for its 24th annual reunion with its paternalistic coach. Adaptation of Miller's Pulitzer Prize winning play is a major, boring disappointment.

FOOTBALL MOVIES

Saints preserve us from another football movie about how all those nice young men are exploited by the big bad colleges! Not that such things aren't all too common; it's just that we've had so many bad movies about it.

If you want to see how to make a good football movie, rent *Brian's Song*, *North Dallas Forty*, or even that old favorite *Knute Rockne: All-American*. For comedy, try *The Longest Yard*, Harold Lloyd, or the Marx Brothers.

THE ALL-AMERICAN *½
(1953) B&W 83 min.
(*Dir.*: Jesse Hibbs. *Cast*: Tony Curtis, Lori Nelson, Richard Long, Mamie Van Doren, Gregg Palmer, Stuart Whitman

Romance 'twixt then-teenage heartthrob Curtis and Nelson is only reason for this picture. Not much football; not much of a movie.

ALL THE RIGHT MOVES **½
(1983) [R] C 91 min. +++
Dir.: Michael Chapman. *Cast*: Tom Cruise, Craig T. Nelson, Lea Thompson, Charles Cioffi.

Good entry in the clichéd "look what win-at-all-costs mentality did to this nice boy" school of sports criticism. Cruise is the young high school star, and Nelson, who stars on TV as an amiable football loser, is much less amiable here as Cruise's hot-headed coach.

THE BEST OF TIMES **
(1986) [PG-13] C 106 min. +++
Dir.: Roger Spottiswoode. *Cast*: Robin Williams, Kurt Russell, Pamela Reed, Holly Palance, Donald Moffat, Margaret Whitton, M. Emmett Walsh, Donovan Scott, R.G. Armstrong.

Williams' life was ruined when he dropped the winning pass in his big high school game. Twenty years later, he talks everybody into a rematch. Uneven comedy has its moments, but eventually fails.

BRIAN'S SONG ****
(1970) C 73 min. TVM +++
Dir.: Buzz Kulik. *Cast*: Billy Dee Williams, James Caan, Jack Warden, Judy Pace, Shelly Fabares.
   Exceptional telling of the real life relationship between Chicago Bears players Gale Sayers (Williams) and Brian Piccolo (Caan). Rivals for the same position, Piccolo helped Sayers rehab from a potentially career-ending injury and then was stricken himself with terminal cancer. Although this deals with death, it is a life-celebrating triumph. One of the best made-for-TV movies *and* one of the best sports-related movies ever.

CRAZYLEGS **
(1953) B&W 87 min.
Dir.: Francis D. Lyon. *Cast*: Elroy Hirsch, Lloyd Nolan, Joan Vohs, Louise Lorimer.
   Tolerable biography of "Crazylegs" Hirsch, who plays himself. In fact, he was good enough to get a "straight" lead in the prison picture *Unchained*, where he was very good again. Then he went back to playing football. For trivia buffs: the song "Unchained Melody" which was resurrected in the movie "Ghost" was introduced in Hirsch's second movie.

EASY LIVING **½
(1949) B&W 77 min. +++
Dir.: Jacques Tourneur. *Cast*: Victor Mature, Lizabeth Scott, Lucille Ball, Sonny Tufts, Lloyd Nolan, Paul Stewart, Jeff Donnell, Jack Paar, Art Baker
   Better-than-you-might-think story of aging footballer pushed to the brink by his grasping wife. Advertised as featuring the Los Angeles Rams, but you'll have to look quick at the background.

EVERYBODY'S ALL-AMERICAN **½
(1988) [R] C 127 min. +++
Dir.: Taylor Hackford. *Cast*: Jessica Lange, Dennis Quaid, Timothy Hutton, John Goodman, Carl Lumbly
   When the football hero marries the homecoming queen, they don't live happily ever after. Quaid and Lange come a-cropper when he ages out of football. This should have been better than it is.

FATHER WAS A FULLBACK **
(1949) B&W 77 min.
Dir.: John M. Stahl. *Cast*: Fred MacMurray, Maureen O'Hara, Betty Lynn, Rudy Vallee, Thelma Ritter, Natalie Wood

Will father's team win the big game? Will all go well on the domestic scene? Will they ever make movies like this again? Not likely. But it's pleasant enough to watch instead of cleaning your attic.

FIGHTING BACK **½
(1980) C 100 min. TVM
Dir.: Robert Lieberman. *Cast*: Robert Ulrich, Art Carney, Bonnie Bedelia, Richard Herd, Howard Cosell, Bubba Smith.

Would you believe there were *three* movies with this title released between 1980-82? This one is the story of how Rocky Bleier of the Pittsburgh Steelers overcame crippling war injuries to become a star NFL running back.

THE FRESHMAN ***
(1925) B&W 70 min.
Dir.: Sam Taylor and Fred Newmeyer. *Cast*: Harold Lloyd, Jobyna Ralston, Brooks Benedict.

There's a good 1990 film of the same name that has nothing to do with football. This *Freshman* is the great silent comedy that has Lloyd as the college patsy who becomes a hero by winning the big game. The magnificent game sequence is reprised in Lloyd's 1947 *The Sins of Harold Diddlebock*, which was then re-edited and released as *Mad Wednesday* in 1950.

GRAMBLING'S WHITE TIGER **½
(1981) C 100 min. TVM +++
Dir.: Georg Stanford Brown. *Cast*: Bruce Jenner, Harry Belafonte, LeVar Burton, Dennis Haysbert, Deborah Pratt, Ray Vitte.

True story of Jim Gregory, who became the only white player on Grambling's football squad, makes a passable TV-movie. Olympian Jenner is a little old for the part, but Belafonte is fine as legendary coach Eddie Robinson.

HORSE FEATHERS ***
(1932) B&W 68 min. +++
Dir.: Norman Z. McLeod. *Cast*: The Marx Brothers, Thelma Todd,

David Landau, Robert Grieg.

The Marx Brothers bring their patented nonsense to a college campus where Groucho is the president, Zeppo is his athlete son, Chico and Harpo are mistaken for football stars. The "game" that furnishes the climax is inspired lunacy. Not really football but irresistible.

THE IRON MAJOR **½
(1943) B&W 85 min.
Dir.: Rex Enright. *Cast*: Pat O'Brien, Ruth Warrick, Robert Ryan.

O'Brien (a.k.a. Knute Rockne) is Frank Cavanaugh, a coach who became a World War I hero in this fairly interesting bio-pic.

JIM THORPE: ALL-AMERICAN **
(1951) B&W 107 min.
Dir.: Michael Curtiz. *Cast*: Burt Lancaster, Charles Bickford, Steve Cochran, Phyllis Thaxter, Dick Wesson.

This could go under "Olympics" because the loss of Thorpe's medals because he'd played some pro baseball is the central incident. However, there's plenty of football—just not much drama.

KNUTE ROCKNE: ALL-AMERICAN ***
(1940) B&W & C 96 min. +++
Dir.: Lloyd Bacon. *Cast*: Pat O'Brien, Ronald Reagan, Gale Page, Donald Crisp, John Qualen.

Corny but entertaining biography of Notre Dame's famed coach. Yes, Reagan asks him to "win one for the Gipper." Whether George Gipp ever asked is another question.

THE LONGEST YARD ***
(1974) [R] C 123 min. +++
Dir.: Robert Aldrich. *Cast*: Burt Reynolds, Eddie Albert, Ed Lauter, Michael Conrad, Jim Hampton, Bernadette Peters, Charles Tyner, Mike Henry, Harry Caesar, Richard Kiel.

Hilarious prison comedy in which Reynolds is forced to put together a team of convicts to play the warden's picked guard team. He comes up with the dirtiest players who ever drew 15 yards for unnecessary mayhem. The last third of the movie is the set-the-audience-cheering, slam-bang game.

NORTH DALLAS FORTY ****
(1979) [R] C 119 +++
Dir.: Ted Kotcheff. *Cast*: Nick Nolte, Mac Davis, Charles Durning, Dayle Haddon, G.D. Spradlin, Bo Svenson, Steve Forrest, John Matuszak, Dabney Coleman.

Pete Gent's novel makes one of the two best football films ever. (The other is the very different "Brian's Song.") A seriocomic account of life and hard times in the NFL with Nolte excellent as an exploited wide receiver and Davis quite good as a Don-Meredith-type quarterback.

NUMBER ONE *
(1969) [M/PG] C 105 min.
Dir.: Tom Gries. *Cast*: Charlton Heston, Jessica Walter, Bruce Dern, John Randolph, Diana Muldaur.

Not even Moses could lead this mess to a higher rating. Heston is an aging quarterback trying to hang on for one more year. You know the drill.

THE ROSE BOWL STORY *
(1952) C 73 min.
Dir.: William Beaudine. *Cast*: Marshall Thompson, Vera Miles, Natalie Wood, Ann Doran, Jim Backus.

There's nothing to recommend here.

SATURDAY'S HERO **
(1951) B&W 111 min.
Dir.: David Miller. *Cast*: John Derek, Donna Reed, Sidney Blackmer, Alexander Knox.

This indictment of big-time college football excesses had more bite 40 years ago.

SATURDAY'S HEROES **
(1937) B&W 60 min.
Dir.: Edward Killy. *Cast*: Van Heflin, Marian Marsh, Richard Lane.

Pretty much the same comment as previous movie but add another decade.

SEMI-TOUGH ***
(1977) [R] C 108 min. +++
Dir.: Michael Ritchie. *Cast*: Burt Reynolds, Kris Kristofferson, Jill

Clayburgh, Robert Preston, Bert Convy, Lotte Lenya, Roger E. Mosley, Richard Masur, Carl Weathers, Brian Denehy, Ron Silver.

Enjoyable comedy about two football players and their mutual girlfriend, based on Dan Jenkins' best-seller. A good assortment of laughs, Reynolds is charming, and the football action is actually believable. With a little more edge, this might have been a semi-classic.

THE SPIRIT OF WEST POINT **
(1947) B&W 77 min. +++
Dir.: Ralph Murphy. *Cast*: Doc Blanchard, Glenn Davis, Tom Harmon, Robert Shayne, Anne Nagel, Alan Hale, Jr.

Army All-Americans Blanchard and Davis play themselves in this account of their glorious days at West Point, and do it fairly well.

THAT'S MY BOY *½
(1951) B&W 98 min.
Dir.: Hal Walker. *Cast*: Dean Martin, Jerry Lewis, Eddie Mayehoff, Marion Marshall, Ruth Hussey, Polly Bergen, John McIntire.

Ex-footballer Mayhoffe wants nerdy son Lewis to follow in his footsteps so he hires Martin (!) to coach him. Adequate for M&L fans.

TROUBLE ALONG THE WAY **
(1953) B&W 110 min.
Dir.: Michael Curtiz. *Cast*: John Wayne, Donna Reed, Charles Coburn, Sherry Jackson, Marie Windsoe, Tom Tully, Leif Erickson, Chuck Connors.

Not one of the Duke's best. He's a divorced football coach trying to keep custody of his daughter and win a few games for a small Catholic school. Sentimentally so-so.

WHAT PRICE VICTORY *
(1988) C 100 min. TVM
Dir.: Kevin Connor. *Cast*: Mac Davis, George Kennedy, Robert Culp, Susan Hess, Guy Boyd, Warren Berlinger, Eric La Salle.

Let's wring our hands over unscrupulous college recruiting. (It'll be more interesting than this dull movie.)

Hockey Movies

Hockey and soccer have never been biggies as far as Hollywood is concerned. Apparently the assumption is that only fans of those two sports will sit through movies with hockey or soccer backgrounds and there aren't enough of either kind of fans out there to make a profit.

THE DEADLIEST SEASON ***
(1977) C 98 min. TVM
Dir.: Robert Markowitz. *Cast*: Michael Moriarty, Kevin Conway, Meryl Streep, Sully Boyar, Jill Eikenberry, Walter McGinn, Andrew Duggan, Patrick O'Neal, Mason Adams.

Thought-provoking story of a defenseman who pleases fans and owners by becoming a "goon." When he accidentally kills another player, he is tried for murder, but who really should be on trial?

SLAP SHOT ***
(1977) [R] C 122 min. +++
Dir.: George Roy Hill. *Cast*: Paul Newman, Michael Ontkean, Lindsay Crouse, Jennifer Warren, Melinda Dillon, Strother Martin, Jerry Houser, Swoosie Kurtz, M. Emmet Walsh, Kathryn Walker, Paul Dooley.

The other side of *The Deadliest Season*—Newman's bush league hockey team becomes successful when they start playing dirty. Plenty of laughs, but the language is not for virginal ears.

YOUNGBLOOD *
(1986) [R] C 109 min. +++
Dir.: Peter Markle. *Cast*: Rob Lowe, Cynthia Gibb, Patrick Swayze, Ed Lauter, Eric Nesterenko, George Finn, Fionnula Flanagan.

More like tired blood. Lowe joins a small-time hockey team, falls in love with the coach's daughter. Ho-hum.

Soccer Movies

VICTORY *½
(1981) [PG] C 110 min. +++
Dir.: John Huston. *Cast*: Sylvester Stallone, Michael Caine, Max von Sydow, Pélé, Daniel Massey, Carole Laure.

Misnamed; this one's a loser. If you can believe a team of Allied POWs would skip a chance to escape their German captors just to finish a soccer game, your faith in the spirit of competition exceeds a rooster's faith in sunrise. The only saving grace is soccer great Pélé's circus kicks.

Boxing Movies

The "boxing movie" has been such a Hollywood staple for so long that this list could be almost endless. We decided to skip some of the lesser-known examples of the '30s and '40s and concentrate on those films that are either already on videotape or likely to be in the next few years.

BODY AND SOUL ***½
(1947) B&W 104 min. +++
Dir.: Robert Rossen. *Cast*: John Garfield, Lilli Palmer, Hazel Brooks, Anne Revere, William Conrad, Joseph Pevney, Canada Lee.

Outstanding film and a real breakthrough for its time. Garfield fights his way to the top using fair means and foul. Gritty, superbly photographed, and strongly acted. Some rank it as a better boxing movie than "Raging Bull," which is saying something.

BODY AND SOUL **
(1981) [R] C 100 min. +++
Dir.: George Bowers. *Cast*: Leon Isaac Kennedy, Jayne Kennedy, Peter Lawford, Michael V. Gazzo, Perry Lang, Kim Hamilton.

Earnest but uninspired remake pales by comparison.

THE CHAMP **½
(1931) B&W 87 min. +++
Dir.: King Vidor. *Cast*: Wallace Beery, Jackie Cooper, Irene Rich, Roscoe Ates.

Sentimental tearjerker (but it works) about a washed-up fighter and his adoring son won Beery an Oscar.

THE CHAMP
(1979) [PG] C 121 min. +++
Dir.: Franco Zeffirelli. *Cast*: John Voight, Faye Dunaway, Ricky Schroeder, Jack Warden, Arthur Hill, Strother Martin.

Awful remake of 1931 film. The original got a lot of popcorn soggy. Here they kept the corn and threw away the pop.

CHAMPION ***½
(1949) B&W, CZ 99 min. +++
Dir.: Mark Robson. *Cast*: Kirk Douglas, Marilyn Maxwell, Arthur Kennedy, Ruth Roman, Paul Stewart, Lola Albright.

The movie that made Douglas a star. He plays an unscrupulous boxer who uses and discards everyone in his climb to the top.

FAT CITY ***
(1972) [PG] C 100 min. +++
Dir.: John Huston. *Cast*: Stacy Keach, Jeff Bridges, Susan Tyrell, Candy Clark, Nicholas Colasanto.

Often underrated little gem about journeyman boxers eking out a precarious living in tank towns. An excellent piece of Americana.

FLESH AND BLOOD **½
(1979) C 200 min. TVM
Dir.: Jud Taylor. *Cast*: Tom Berenger, Mitchell Ryan, Kristin Griffith, Denzel Washington, Suzanne Pleshette, John Cassavetes, Bert Remsen, Dolph Sweet.

Story of boxer's rise and fall is okay, but it's been done better.

GOLDEN BOY **½
(1939) B&W 99 min. +++
Dir.: Rouben Mamoulian. *Cast*: William Holden, Barbara Stanwyck, Adolphe Menjou, Lee J. Cobb, Joseph Calleia, Sam Levene.

Based on Clifford Odets' play, this story of fighter who'd rather be a musician has aged badly. Holden's first film.

THE GOLDEN GLOVES STORY *
(1949) B&W 76 min.
Dir.: Felix E. Feist. *Cast*: James Dunn, Dewey Martin, Kay Westfall, Kevin O'Morrison.

The effect of an upcoming bout on two young boxers' lives. Nothing special.

THE GREAT WHITE HOPE ***
(1970) [PG] C 101 min. +++
Dir.: Martin Ritt. *Cast*: James Earl Jones, Jane Alexander, Lou Gilbert, Joel Fluellen, Chester Morris, Robert Webber, R.G. Armstrong, Hal Holbrook, Moses Gunn.

Jones and Alexander repeat their roles from the hit stage production in this excellent social and character study. The subject is Jack Johnson (called Jefferson here) and the white reaction when a black man first held the heavyweight crown.

THE HARDER THEY FALL ***½
(1956) B&W 109 min. +++
Dir.: Mark Robson. *Cast*: Humphrey Bogart, Rod Steiger, Jan Sterling, Mike Lane, Max Baer, Edward Andrews.

Excellent Budd Schulberg script casts Bogart (in his last feature film) as a cynical sportswriter who becomes a press agent for a young and naive, Primo Carnera-type heavyweight. He sees for the first time the corruption, sleaze and manipulation from the inside but can't save his man from a horrendous beating.

HEART OF A CHAMPION: THE RAY MANCINI STORY **
(1985) C 100 min. TVM +++
Dir.: Richard Michaels. *Cast*: Robert Blake, Doug McKeon, Mariclare Costello, Tony Burton.

True (but not particularly interesting) tale of lightweight champ "Boom-Boom" Mancini who won title his father couldn't because of World War II service.

IRON MAN **
(1951) B&W 82 min.
Dir.: Joseph Pevney. *Cast*: Jeff Chandler, Evelyn Keyes, Stephen McNally, Rock Hudson, Joyce Holden, Jim Backus.

Unhappy rise of iron-jawed boxer with a gold-digging wife.

JACK JOHNSON ***
(1971) [PG] B&W 90 min.
Dir.: William Cayton.
  Documentary about life of first black champion. Narrated by Brock Peters, with a Miles Davis score.

THE JOE LOUIS STORY **
(1953) B&W 88 min. +++
Dir.: Robert Gordon. *Cast*: Coley Wallace, Paul Stewart, Hilda Simms, James Edwards.
  Not the worst bio-pic ever. Not the best.

THE KID FROM BROOKLYN **½
(1946) C 113 min. +++
Dir.: Norman Z. McLeod. *Cast*: Danny Kaye, Virginia Mayo, Vera Ellen, Steve Cochran, Eve Arden, Walter Abel.
  Kaye vehicle about a milkman turned prizefighter has some funny moments.

THE PRIZE FIGHTER *½
(1979) [PG] C 99 min. +++
Dir.: Michael Preece. *Cast*: Tim Conway, Don Knotts, David Wayne.
  Dull comedy about dumb boxer and smart-aleck manager is strictly for the kids. Very young kids.

RAGING BULL ****
(1980) [R] B&W 128 min. +++
Dir.: Martin Scorsese. *Cast*: Robert De Niro, Cathy Moriarty, Joe Pesci, Frank Vincent, Nicholas Colasanto, Theresa Saldana.
  Brilliant work by director Scorsese and star De Niro (Oscar winner) results in a great film. The untidy life of ex-champ Jake LaMotta, whose greatest enemy was himself, from his early 20s to his 40s makes an unlikely but riveting story.

REQUIEM FOR A HEAVYWEIGHT ***
(1965) B&W 100 min.
Dir.: Ralph Nelson. *Cast*: Anthony Quinn, Jackie Gleason, Mickey Rooney, Julie Harris, Nancy Cushman, Madame Spivy, Cassius Clay (Muhammad Ali).
  Rod Serling script adapted from his highly successful TV drama casts Quinn as a washed-up fighter whose search for a new career

leads him into degradation. Rooney as friend and Harris as social worker stand out among a uniformly excellent cast.

ROCKY ***
(1976) [PG] C 119 min. +++
Dir.: John G. Avildsen. *Cast*: Sylvester Stallone, Talia Shire, Burt Young, Carl Weathers, Burgess Meredith.

This story of a bum boxer getting his one shot at fame and fortune was voted Best Picture, which was a bit of a reach. Still, the bang-up climax had everybody on their feet cheering.

ROCKY II **½
(1979) [PG] C 119 min. +++
Dir.: Sylvester Stallone. *Cast*: Sylvester Stallone, Talia Shire, Burt Young, Carl Weathers, Burgess Meredith.

Much the same movie as I, with another terrific ending to save what is essentially a 1930s B-movie.

ROCKY III **
(1982) [PG] C 99 min. +++
Dir.: Sylvester Stallone. *Cast*: Sylvester Stallone, Talia Shire, Burt Young, Carl Weathers, Burgess Meredith, Mr. T.

This is the one where he fights Mr. T, Fool!

ROCKY IV *½
(1985) [PG] C 91 min. +++
Dir.: Sylvester Stallone. *Cast*: Sylvester Stallone, Dolph Lundgren, Carl Weathers, Burt Young, Talia Shire.

Here Rocky fights the Russian for World Peace.

ROCKY V *
(1990) [PG-13] C 104 min. +++
Dir.: John D. Avildsen. *Cast*: Sylvester Stallone, Talia Shire, Burt Young, Sage Stallone, Burgess Meredith.

In the fifth installment, Rocky tries to come back even though he's brain-damaged from his last film—er, bout. So he's really fighting himself, see? What next? Rocky vs. Rambo? Rocky vs. Bullwinkle?

THE SET-UP ***½
(1949) B&W 72 min. +++
Dir.: Robert Wise. *Cast*: Robert Ryan, Audrey Totter, George

Tobias, Alan Baxter, James Edwards, Wallace Ford.

Strong little film about a washed-up boxer (Ryan) who refuses to take a dive for shady promoters. Sense of urgency is enhanced by playing movie in real time—72 minutes in a man's life.

SOMEBODY UP THERE LIKES ME ***
(1956) B&W 113 min. +++
Dir.: Robert Wise. *Cast*: Paul Newman, Pier Angeli, Everett Sloane, Eileen Heckart, Sal Mineo, Joseph Buloff, Robert Loggia, Steve McQueen.

Biography of Rocky Graziano still stands up well. Fine performance by Newman.

TRIUMPH OF THE SPIRIT **½
(1989) [R] C 121 min. +++
Dir.: Robert M. Young. *Cast*: Willem Dafoe, Edward James Olmos, Robert Loggia, Wendy Gazelle.

Heart-wrenching, true story of Greek-Jewish boxer sent to Auschwitz who survives in the camp by boxing. Grim, but often compelling.

WORLD IN MY CORNER **
(1956) B&W 104 min.
Dir.: Jesse Hibbs. *Cast*: Audie Murphy, Barbara Rush, Jeff Morrow.

Poor boy rises in boxing and is nearly ruined by fame. You've seen it before, done better.

Golf Movies

DEAD SOLID PERFECT **½
(1988) C 95 min. TVM +++
Dir.: Bobby Roth. *Cast*: Randy Quaid, Kathryn Harrold, Jack Warden, Corinne Nohrer.

Made-for-cable comedy-drama based on Dan Jenkins' novel about life and loves on the pro tour. Title sounds like a murder mystery but it refers to when a golfer hits the ball perfectly. Not great, but worth your time.

FOLLOW THE SUN ∗∗
(1951) B&W 93 min.
Dir.: Sidney Lanfield. *Cast*: Glenn Ford, Anne Baxter, Dennis O'Keefe, June Havoc.

Usual Hollywood-hokey biography. Here the victim is golfing great Ben Hogan.

PAT AND MIKE ∗∗∗
(1952) B&W 92 min. +++
Dir.: George Cukor. *Cast*: Spencer Tracy, Katharine Hepburn, Aldo Ray, William Ching, Jim Backus, Chuck Connors.

The Dynamic Duo couldn't make a bad picture. This isn't their best but it's still a "cherce" comedy. She's a top woman athlete and he's her manager. Aldo Ray scores as dumb boxer.

Skiing Movies

Honestly, our favorite ski scene is when Gregory Peck and Ingrid Bergman head for the cliff in *Spellbound*. Our second faves are any one of several downhill chases in James Bond flicks. Then there are several movies where skiing is just an excuse to show snow bunnies cavorting in ways you'd never tell your mother about. As for actual "ski movies," the pickings are slim.

DOWNHILL RACER ∗∗∗
(1969) [M/PG] C 102 min. +++
Dir.: Michael Ritchie. *Cast*: Robert Redford, Gene Hackman, Camille Spary, Karl Michael Vogler, Dabney Coleman.

Redford is excellent as a self-centered egotist on the Olympic ski team. On the other hand, it's hard to care whether he skis or falls off his next mountain. Dazzling ski footage.

GOING FOR THE GOLD: THE BILL JOHNSON STORY ∗∗
(1985) C 100 min. TVM +++
Dir.: Don Taylor. *Cast*: Anthony Edwards, Dennis Weaver, Sarah Jessica Parker.

From punk to ski-hunk, ordinary telling of Johnson's rise to Olympics competition.

OTHER SIDE OF THE MOUNTAIN **½
(1975) [PG] C 101 min. +++
Dir.: Larry Peerce. *Cast*: Marilyn Hassett, Beau Bridges, Dabney Coleman.
    Okay but not outstanding true-life story of Olympic skier Jill Kinmont and her gallant fight back from paralyzing ski accident.

Tennis Movies

Did you ever notice that the hero of Hitchcock's 1951 *Strangers on a Train* and the villain of his 1954 *Dial M for Murder* were both tennis players? Was he trying to tell us something about his changing attitude toward the game?

JOCKS *
(1987) [R] C 91 min. +++
Dir.: Steve Carter. *Cast*: Scott Strader, Perry Lang, Mariska Hargitay, Richard Roundtree, Christopher Lee, R.G. Armstrong.
    Smirky sex comedy about a college tennis team's fun and games in a Las Vegas tournament.

LITTLE MO **½
(1978) C 150 min. TVM +++
Dir.: Daniel Haller. *Cast*: Glynnis O'Connor, Michael Learned, Anne Baxter, Claude Akins, Martin Milner, Anne Francis, Leslie Nielsen.
    TV bio-pic about teenage tennis star Maureen Connolly and her tragic battle with cancer.

Billiards (Pool) Movies

THE HUSTLER ****
(1961) B&W 135 min. +++
Dir.: Robert Rossen. *Cast*: Paul Newman, Jackie Gleason, Piper

Laurie, George C. Scott, Myron McCormick, Murray Hamilton.

One of Newman's best roles as Fast Eddie Felson, the pool hall hustler and (maybe) Loser, who challenges Minnesota Fats (Gleason). Great cast, atmosphere, suspense.

THE COLOR OF MONEY ***
(1986) [R] C 119 min. +++
Dir.: Martin Scorsese. *Cast*: Paul Newman, Tom Cruise, Mary Elizabeth Mastrantonio, Helen Shaver, John Turturro.

After twenty-six years, the sequel to "The Hustler." Though not up to the original (and the ending is a hanger) this is a very good movie. Cruise is excellent as a hotshot young protégé, and Newman finally got the Oscar he (maybe) should have won the first time around.

Horse Racing Movies

There are far too many girl/boy and his/her horse stories that could be thought of as racing sagas to include here. Nor do we address gambling fever, race-fixing thrillers, or talking animal stories. We'll just recommend the following as well worth your time:

CHAMPIONS ***
(1984) [PG] C 115 min. +++
Dir.: John Irvin. *Cast*: John Hurt, Edward Woodward, Jan Francis, Ben Johnson, Kirstie Alley.

True story of English steeplechase jockey Bob Champion (Hurt) who fought and won over cancer to win the 1981 Grand National.

PHAR LAP ***
(1983) [PG} C 108 min. +++
Dir.: Simon Wincer. *Cast*: Tom Burlinson, Martin Vaughan, Ron Liebman, Celia de Burgh.

Entertaining Australian film about the legendary racehorse from Down Under. Burlinson is the stable boy who first saw greatness in the horse.

Track and Field/Olympics Movies

BABE ***
(1975) C 100 min. TVM
Dir.: Buzz Kulik. *Cast*: Susan Clark, Alex Karras, Slim Pickens, Jeanette Nolan, Ellen Geer, Ford Rainey.

Very good bio of superior woman athlete Babe Didrickson Zaharias, with an Emmy-winning performance by Clark in the title role and a fine job by football star Karras as her husband.

THE BOB MATHIAS STORY **½
(1954) B&W 80 min.
Dir.: Francis D. Lyon. *Cast*: Bob Mathias, Ward Bond, Melba Mathias, Paul Bryar, Ann Doran.

Okay telling of the life of the two-time Olympic decathlon champ, with Mathias doing a good turn as himself.

CHARIOTS OF FIRE ****
(1981) [PG] C 123 min. +++
Dir.: Hugh Hudson. *Cast*: Ben Cross, Ian Charleson, Nigel Havers, Nack Farrell, Alice Krige, Cheryl Campbell, Ian Holm, John Gielgud.

Wonderful movie about two men who competed in the 1924 Olympics, exploring their motives, psychology, emotions. Won Oscar as best picture.

THE JESSE OWENS STORY ***
(1984) C 200 min. TVM +++
Dir.: Richard Irving. *Cast*: Dorian Harewood, Georg Stanford Brown, Debbi Morgan, Tom Bosley, LeVar Burton, Ronny Cox, Ben Vereen, Greg Morris, George Kennedy.

Fine bio of the man who won four gold medals in the 1936 Olympics. Follows him from his great success to his shameless exploitation afterward. Harewood is exceptional in the title role.

THE LONELINESS OF THE LONG DISTANCE RUNNER ****
(1962) B&W 103 min.
Dir.: Tony Richardson. *Cast*: Michael Redgrave, Tom Courtenay,

Avis Brunnage, Peter Madden, Alec McCowen, James Fox.

Great film about the rebellion of a youth chosen to represent his school in a distance race. Not really a sports picture, but so good we couldn't leave it out.

NADIA **
(1984) C 100 min. +++
Dir.: Alan Cooke. *Cast*: Talia Balsam, Jonathan Banks, Joe Bennett, Simone Blue, Johann Carlo, Conchita Ferrell, Carrie Snodgrass.

Nadia Comaneci electrified the world with her gymnastics performance in the 1976 Olympics, but this bland telling of her story can't light up a small room. At its best in the gym.

PERSONAL BEST ***
(1982) [R] C 124 min. +++
Dir.: Robert Towne. *Cast*: Mariel Hemingway, Scott Glenn, Patrice Donnelly, Kenny Moore, Jim Moody, Larry Pennell.

Although the lesbian relationship between the two women athletes portrayed by Hemingway and Donnelly probably sold theater tickets, it's secondary to this gripping story of athletes in training for the 1980 Olympics.

SEE HOW SHE RUNS ***
(1978) C 100 min. TVM +++
Dir.: Richard T. Heffron. *Cast*: Joanne Woodward, John Considine, Lissy Newman, Mary Beth Manning, Barnard Hughes.

Woodward is a middle-aged housewife who decides to turn her life around by entering the grueling Boston Marathon against nearly everybody's advice. You'll be there cheering at the finish line. Woodward won a much-deserved Emmy.

INDEXES

Baseball

Aaron, Hank, 48
Amazin' Era: New York Mets 23 Years, 43
Art of Hitting, 51
Art of Hitting .300, 51

Ball Talk: Baseball's Voices of Summer, 50
Baseball 1968, 38
Baseball 1969, 38
Baseball Bunch, 52
Baseball Bunch: Hitting, 52
Baseball Bunch: Pitching, 52
Baseball Card Collector, 50
Baseball Dream Team (American League), 50
Baseball Dream Team (National League), 50
Baseball Dynasties: The New York Yankees, Oakland A's and Cincinnati Reds, 50
Baseball Fun and Games, 56
Baseball Funnies: A Hilarious Look at Baseball, 56
Baseball Funny Side Up, 57
Baseball In the '70s, 38
Baseball In the '80s, 40, 47
Baseball In the News, 37
Baseball Laughs, Gaffes & Goofs, 57
Baseball Legends, 50
Baseball Masters Series, 52
Baseball News Highlights 1959, 37
Baseball Our Way, 52
Baseball Rivalries: The Yankees and the Dodgers, 51
Baseball Skills and Drills With Dr. Bragg Stockton, 52
Baseball the Pete Rose Way, 53
Baseball the Right Way, 53
Baseball the Yankee Way, 53
Baseball Time Capsule: A Journey Through the Barry Halper Collection, 50
Baseball Tips, 53
Baseball with Rod Carew, 52
Baseball's Greatest Hits, 50
Baseball's Greatest Moments, 36
Baseball's Official Ballpark Bloopers, 57
Baseball's Record Breakers, 36
Baserunning Basics with Maury Wills, 52
Battlin' Bucs: The First 100 Years of the Pittsburgh Pirates, 46
Bench, Johnny, 48
Berra, Yogi, 48
Boys of Summer, The, 42
Brett's Secrets of Baseball, George, 53

Canseco's Baseball Camp, Jose, 53
Catching, Bunting, Outfield Play, Baserunning and Sliding, 54
Catching with Lance Parrish, 55

Centennial: Over 100 Years of Philadelphia Phillies Baseball, 45
Chicago and the Cubs: A Lifelong Love Affair, 41
Chicago White Sox: A Visual History, 41
Chicago's Grand Stands: Chicago's Classic Ballpark, 41
Clemente: A Touch of Royalty, Roberto, 48
Coaching Clinic, 53
Coaching Psychology, 52
Conditioning and Baserunning, 52

Dawson: He's a Hero, Andre, 48
Decade of Transition: The '70s, 38
Defensive Skills by Position, 50
Detroit Tigers: The Movie, 42
Diamond in the Emerald City: 10 Years of Seattle Mariners Baseball, 47
DiMaggio, Joe, 48
Do It Better, 53
Doctor's Prescription for the Pitcher: A Step-by-Step Total Body Conditioning Program Medically Designed to Improve Pitching Performance, 53
Dodger Stadium: 25th Anniversary, 42
Dodgers Way to Play Baseball, 53
Drysdale, Don, 48
Dynasty: The New York Yankees, 44

Fantastic Baseball Bloopers, 57
Fenway: 75 Years of Red Sox Baseball, 40
Fielding, 52
Fielding for Kids, 53
Ford, Whitey, 48
Fundamentals of Baserunning, 52
Fundamentals of Fielding, 52
Fundamentals of Hitting, 52
Fundamentals of Pitching, 52
Future Legends of Baseball, 51

Game of the Week: Yankees vs. Washington Senators, 44
Garvey, Steve, 48
Garvey's Hitting System, Steve, 56
Gehrig, Lou, 48

Giants History: The Tale of Two Cities, 46
Glory of Their Times, The, 36
Golden Decade of Baseball 1947–1957, 37
Golden Greats of Baseball, 51
Golden Greats of Baseball: Pitchers, 51
Grand Slam, 51
Great World Series Heroes: The Men Who Made it Happen, 34
Greatest Home Run Hitters, 51
Guidry's 18 Strikeout Performance, Ron, 44
Gwynn's King of Swing, Tony, 53
Gwynn's Play to Win, Tony, 53

History of Baseball: Greatest Moments From Baseball's Past, 37
History of Great Black Baseball Players, 37
Hitting, 52
Hitting for Kids, 53
Hitting Machine: Featuring the Star System (Stride, Trigger, Assemble, Release), 54
Hitting with Al Kaline, 55
Hitting with Harry "The Hat" Walker, 54
Hitting: Getting the Feel of It, 54
How to Play Better Baseball, 54
Howser's Baseball Workout, Dick, 54

Infield Techniques and Catching with Bud Middaugh, 54
Introduction to Baseball Card Collecting, 51

Jackson, Reggie, 48
Jackson: Mr. October, Reggie, 48

Legend of Stan the Man Musial, 49
Little League: How to Hit and Run, 54
Little League: How to Pitch and Field, 54
Little League's Official How-to-Play Baseball by Video, 54

M.V.P.: World Series Edition, 34

Martin: The Man, the Myth, the Manager, Billy, 49
Master the Secrets of the Hitting Machine, 54
Meet Babe Ruth, 49
Mantle, Mickey, 48
Mantle, Mickey/Willie Mays, 49
Mantle: The American Dream Comes to Life, Mickey, 48
Mantle's Baseball Tips for Kids of All Ages, Mickey, 54
Mays, Willie, 49
Mission Impossible: The First Decade—10 Years of Toronto Blue Jays Baseball, 47
Musial, Stan, 49
My Dad, the Babe: The Babe Ruth Scrapbook, 49

New Era, 44
New York Yankees Great Game Broadcasts, 44
New York Yankees Special Theme Cassettes, 44
New York Yankees: The Movie, 44

Oakland A's: All Star Almanac, 45
October Spotlight: World Series Heroes: The Men Who Made It Happen, 34
Once in a Lifetime: World Series Heroes, 34

Palmer, Jim, 49
Piniella's Winning Ways, Lou, 55
Pinstripe Power: Story of the 1961 Yankees, 44
Pitcher's Fielding Practice, 55
Pitching, 52
Pitching Absolutes, 55
Pitching For Kids, 53
Pitching, Hitting and Infield Play, 54
Pitching Mechanics: Problem Recognition and Solutions, 55
Pitching Strategies and Tactics, 55
Pitching with Bud Middaugh, 55
Pitching with Roger Craig, 55
Play Ball Series, 55
Play Ball with Mickey Mantle, 55
Play Ball with Reggie Jackson, 55
Play Ball with the Yankees, 44

Pro Baseball's Funniest Pranks, 57
Professional Hitter, 55
Professional Sports Training for Kids, 55

Reds: The Official History of the Cincinnati Reds, 41
Righetti's No-hit Game, Dave, 44
Robinson, Brooks, 49
Robinson, Frank, 49
Robinson, Jackie/Roberto Clemente, 49
Rose, Pete, 49
Rose On Winning Baseball, Pete, 56
Ruth, Babe, 49
Ruth, Babe/Joe DiMaggio, 49
Ruth: The Man, the Myth, the Legend, Babe, 49
Ryan, Nolan, 49
Ryan: Feel the Heat, Nolan, 49

St. Louis Cardinals: The Movie. 109 Years of the Cardinals, 46
Schmidt, Mike, 50
Science of Pitching, 56
Silver Odyssey: 25 Years of Houston Astros Baseball, 42
Simplified Fundamentals of Hitting and Bunting, 56
(Smith): The Movie, Ozzie, 50
Snider, Duke, 50
Sports Clinic: Baseball, 56
Sports Teaching Video: Baseball, 56
Sports Training Camp: Baseball, 56
Super Duper Baseball Bloopers, 57

Teaching Kids Baseball with Jerry Kindall, 56
Teaching The Mechanics of the Major League Swing, 56
Then & Now: The Minnesota Twins' Silver Anniversary (1961–1985), 43
This Week in Baseball 1990, 38
This Week in Baseball's Greatest Play, 38

Umpiring Baseball: The Third Team on the Field, 56

Williams, Ted, 50

Williams, Ted/Pete Rose, 50
Winning Tradition, 45
World Series Unsung Heroes, 34

Yankee Stadium: Home of Heroes, 45
Yankees-Red Sox Playoff Game, 44
Year in Baseball 1988, 38

10 Greatest Moments in Yankee History, 44
100 Years: A Visual History of the Dodgers, 42
1938 American League Film: The First Century of Baseball, 37
1943 World Series, Yankees vs. Cardinals, 29
1944 World Series, Cardinals vs. Browns, 29
1945 World Series, Tigers vs. Cubs, 29, 30
1946 World Series, Cardinals vs. Red Sox, 30
1947 Boston Braves: The Braves Family, 39
1947 World Series, Yankees vs. Dodgers, 30
1948, 1952 & 1955 All Star Games, 34
1948 World Series, Indians vs. Braves, 30
1949 World Series, Yankees vs. Dodgers, 30
1950 World Series, Yankees vs. Phillies, 30
1950s Baseball Dream Team, 51
1951 Baseball News, 37
1951 World Series, Yankees vs. Giants, 30
1952 World Series, Yankees vs. Dodgers, 30
1953 World Series, Yankees vs. Dodgers, 30
1954 Boston Red Sox: Baseball in Boston, 40
1954 Milwaukee Braves: Home of the Braves, 39
1954 World Series, Giants vs. Indians, 30
1955 Boston Red Sox: The Red Sox at Home, 40
1955 Milwaukee Braves: Baseball's Main Street, 39
1955 Washington Senators: Story of the Washington Senators, 37
1955 World Series, Dodgers vs. Yankees, 30
1955–1956 Baseball News, 37
1956 All Star Game, 34
1956 Boston Red Sox: Pride of New England, 40
1956 Kansas City Athletics: The Kansas City A's in Action, 45
1956 Milwaukee Braves: Bravesland, U.S.A, 39
1956 New York Yankees: Winning With the Yankees, 44
1956 World Series, Yankees vs. Dodgers, 31
1957 Boston Red Sox: Play Ball with the Red Sox, 40
1957 Cardinal Tradition and the Game Nobody Saw, 46
1957 Milwaukee Braves: Hail to the Braves!, 39
1957 World Series, Braves vs. Yankees, 31
1958 Detroit Tigers: Tigertown U.S.A. and 1962: Baseball for Little Leaguers, 42
1958 World Series, Yankees vs. Braves, 31
1959 Milwaukee Braves: Fighting Braves of '59, 39
1959 World Series, Dodgers vs. White Sox, 31
1960 Milwaukee Braves: The Best of Baseball, 39
1960 Pittsburgh Pirates: We Had 'Em All the Way, 46
1960 World Series, Pirates vs. Yankees, 31
1961 Minnesota Twins, 43
1961 World Series, Yankees vs. Reds, 31
1962 All Star Game, 34
1962 Baltimore Orioles: The Baltimore Orioles In Action, 39
1962 Los Angeles Angels: Angels '62, 40
1962 World Series, Yankees vs. Giants, 31

1963 World Series, Dodgers vs. Yankees, 31
1964 World Series, Cardinals vs. Yankees, 32
1965 All Star Game, 34
1965 New York Mets: Expressway to the Big Leagues, 43
1965 World Series, Dodgers vs. Twins, 32
1966 All Star Game, 35
1966 Baltimore Orioles, 39
1966 World Series, Orioles vs. Dodgers, 32
1967 All Star Game, 35
1967 New York Mets: Year of Change, 43
1967 World Series, Cardinals vs. Red Sox, 32
1968 World Series, Tigers vs. Cardinals, 32
1969 Baltimore Orioles, 39
1969 Cubs vs. Phillies Complete Game, 41
1969 World Series, Mets vs. Orioles, 32
1970 All Star Game, 35
1970 American League, 38
1970 National League, 38
1970 World Series, Orioles vs. Reds, 32
1971 All Star Game, 35
1971 World Series, Pirates vs. Orioles, 32
1972 All Star Game, 35
1972 World Series, A's vs. Reds, 32
1973 All Star Game, 35
1973 World Series, A's vs. Dodgers, 32
1974 All Star Game, 35
1974 World Series, A's vs. Dodgers, 32
1975 All Star Game, 35
1975 American League, 38
1975 National League, 38
1975 World Series, Reds vs. Red Sox, 33
1976 All Star Game, 35
1976 World Series, Reds vs. Yankees, 33
1977 All Star Game, 35
1977 World Series, Yankees vs. Dodgers, 33
1978 All Star Game, 35
1978 New York Yankees: Greatest Comeback Ever, 44
1978 World Series, Yankees vs. Dodgers, 33
1978: It Don't Come Easy, 44
1979 All Star Game, 35
1979 World Series, Pirates vs. Orioles, 33
1979-81 Baltimore Orioles, 39
1980 All Star Game, 35
1980 Philadelphia Phillies: We Win!, 45
1980 World Series, Phillies vs. Royals, 33
1981 All Star Game, 35
1981 World Series, Dodgers vs. Yankees, 33
1982 All Star Game, 35
1982 Baltimore Orioles, 39
1982 World Series, Cardinals vs. Brewers, 33
1983 All Star Game, 36
1983 Baltimore Orioles, 40
1983 World Series, Orioles vs. Phillies, 33
1984 All Star Game, 36
1984 Baltimore Orioles, 40
1984 World Series, Tigers vs. Padres, 33
1985 All Star Game, 36
1985 Kansas City Royals: Thrill of it All, 42
1985 St. Louis Cardinals: Heck of a Year, 46
1985 World Series, Royals vs. Cardinals, 33
1986 All Star Game, 36
1986 Boston Red Sox: 1986 A.L. Champions, 40
1986 Cleveland Indians: Indian Summer, 41
1986 Milwaukee Brewers: Family Album, 43
1986 Minnesota Twins: Power & Promise, 43
1986 New York Mets: A Year to Remember, 43
1986 Philadelphia Phillies: Headed for the Future, 46

1986 San Francisco Giants: You Gotta Like This Team, 47
1986 Texas Rangers, 47
1986 This Year in Baseball, 38
1986 World Series, Mets vs. Red Sox, 33
1987 All Star Game, 36
1987 Minnesota Twins: Twins Win, 43
1987 St. Louis Cardinals: That's a Winner, 46
1987 World Series, Twins vs. Cardinals, 33
1988 All Star Game, 36
1988 Boston Red Sox: Morgan's Magic, 40
1988 Los Angeles Dodgers: Through the Eyes of a Winner, 42
1988 Oakland A's: A Bashing Success, 45
1988 Phillies' Home Companion, Vol. I: The Game's Easy, Harry, 46
1988 World Series, Dodgers vs. A's, 33
1989 Baltimore Orioles, 40
1989 Chicago Cubs: Boys of Zimmer, 41
1989 Oakland A's and San Francisco Giants: Champions by the Bay, 45, 47
1989 Phillies' Home Companion, Vol. II: Not Necessarily Another Day at the Yard, 46
1989 Toronto Blue Jays: Sky High— A.L. East Champions, 47
1989 World Series, A's vs. Giants, 33
1990 Boston Red Sox: Beating The Odds, 40
1990 Cincinnati Reds: Wire to Wire, 41
1990 New York Mets, The, 43
1990 Oakland Athletics and San Francisco Giants: A Call to Arms, 41, 47
1990 Phillies' Home Companion, Vol. III: Gettin' Dirty, 46
1990 Pittsburgh Pirates: No Doubt About It, 46
1990 Toronto Blue Jays: Tradition of Success, 48
1990 World Series, Reds vs. A's, 34
1991 World Series, Twins vs. Braves, 34
20 Years of World Series Thrills: 1938–1958, 29
50 Years of Little League Baseball, 51
50 Years of Yankee All Stars, 44
500 Home Run Club, 51
75 Years of World Series Memories, 29

Basketball

ACC Basketball: 35 Years of Excellence, 66
Abdul-Jabbar, Kareem, 71
Alford's 50-Minute All-American Workout, Steve, 77
All New Dazzling Dunks and Basketball Bloopers, 78
Arkansas Razorback Attack, 72
Atlanta Hawks 1986–87: Basketball's Air Force, 69
Auerbach, Red, 71

Ball Handling, 73
Barry: Shooting and Offensive Moves, Rick, 76
Basketball Coaches Corner Teaching Tape, 72
Basketball Dream Team, 67
Basketball for the '90s: Beginning Basketball with Bibby, 72
Basketball for the '90s: Building a Championship Defense, 72
Basketball for the '90s: Building a Championship Offense, 72
Basketball for the '90s: Developing the Big Man, 72
Basketball for the '90s: Developing the Perimeter Player, 72
Basketball for the '90s: The "Shark" on Defense, 72
Basketball for the '90s: The "Shark" on Offense, 73
Basketball in the Fast Lane, 72
Basketball Laughs, Gaffes, and Goofs, 78
Basketball with Bill Foster and Gail Goodrich, 73
Basketball with Hubie Brown, 73
Basketball's Amazing Rams, Slams and Jams, 67, 78
Becoming a Basketball Player Series, 73
Becoming a Great Shooter, 74
Big Man Camp, 73
Bird: Basketball Legend, Larry, 71
Boeheim on Basketball, 73
Bombs Away: Let's for 3, 73
Boston Celtics 1986–87: Home of the Brave and Sweet 16, 69

Chamberlain, Wilt, 71
Chamberlain, Wilt / Kareem Abdul-Jabbar 71
Championship Approach to Strength Training and Conditioning, 74
Championship Preparation for Games, 74
Charlotte Hornets 1988–89: Hornet Hysteria, 69
Chicago Bulls 1987–88: Higher Ground, 69
Chicago Bulls 1990–91: Learning to Fly, 69
Coach to Coach, 74

Dallas Mavericks 1987–88: Bouncing Back, 69
Dazzling Dunks and Basketball Bloopers, 78
Defense and Rebounding, 73
Detroit Pistons 1987–88: Bad Boys, 69
Detroit Pistons 1988–89: Motor City Madness, 69
Detroit Pistons 1989–90: Pure Pistons, 69
Developing Your Perimeter Players, 74
Developing Your Post Players, 74
Do it Better: Basketball, 74
Do it Better: Women's Basketball, 74
Dr. J's Basketball Stuff, 74
Drills for Teaching Individual Fundamentals and Team Defense, 74
Drills for Teaching Individual Fundamentals and Team Offense, 74
Dynasty Renewed, The, 67

Erving, Julius, 71

Fantastic Basketball Bloopers, 78
Fast Break: The Fundamentals of Championship Basketball, 75
Final Four: The Movie, 66
Fratello's 3-Point Play, Mike, 75

INDEXES

Golden Greats of Basketball, 68
Great Moments in College Basketball, 66

Harlem Globetrotters: Six Decades of Magic, 68
Havlicek, John, 71
High Percentage Basketball: Getting the Ball to the Open Man, 75
History of the NBA, 68
Hoops Bloops, 79
Houston Rockets 1986–87: Hanging Tough, 69
Hurley's Championship Basketball Series, Bob, 73

Johnson: Always Showtime, Magic, 71
Johnson: Put Magic in Your Game, Magic, 75
Jordan: Come Fly with Me, Michael, 71
Jordan's Playground, Michael, 75

Kareem: Reflections From Inside, 71
Knight's Basketball Clinic Series, Bobby, 73
Knight of Basketball, 75

Legendary Big Men, 68
Los Angeles Lakers 1986–87: World Championship Drive for Five, 70
Los Angeles Lakers 1987–88: Back to Back, 70

Madison Square Garden's All-Time Greatest Basketball Featuring the Harlem Globetrotters, 68
March Through Madness: 1990–91 Kansas Basketball, 66
Massimino Basketball Instructional Tape, Rollie, 77
Match-Up Zone, 74
Maximizing Your Defense: Combatting Today's Offensive Trends, 75
Meyer Basketball Series, Don, 74
Miami Heat 1988–89: The Dream Catches Fire, 70
Minneapolis Lakers 1950–51: Basketball Fundamentals and 1952–53 Meet the Champs, 70

Motion Offenses: Simplified Principles, 74

NBA All-Star Weekend 1986, 68
NBA Awesome Endings, 68
NBA Showmen: The Spectacular Guards, 68
NBA Superstars, 68
New York's Game, 70
North Carolina System, 75

Offensive and Defensive Basketball Techniques, 76
Offensive Low Post Play, 76
Offensive Moves off the Dribble, 73
Offensive Moves, 73
Official NCAA Championship Video: 1990, 67
Official NCAA Championship Video: 1991 67
Olson's Basketball Series, Lute, 75

Pistol Pete's Homework Basketball Series, 76
Pitino Basketball Series, Rick, 77
Point Guard Play, 76
Practice Planning, Organization and On-the-Floor Demonstration, 74
Pressure Defense: A System, 76
Pressure Man-to-Man Defense: A System, 74
Pro Basketball's Funniest Pranks, 79

Reach for the Skies with Spud Webb, 76
Red on Roundball, 76
Reebok Sports Library: Basketball Skills, 76
Robertson, Oscar, 71
Russell, Bill, 71

Shooting to Win: Fundamentals of Shooting by Steve Alford, 77
Shooting, 73
Sports Clinic Basketball, 77
Sports Teaching Video: Basketball, 77
Sports Training Camp: Basketball, 77
Star Shot, 77
Super Slams of the NBA, 68

Tar Heels on Tape, 67
Tar Heels on Tape: The Team That Wouldn't Quit, 67
Teaching Kids Basketball with John Wooden, 77
Team Attitude, 74
Temple of Zones, 77
Time Out, Baby! Dick Vitale's College Hoop Superstars, 67
Transition Game, 74

Unknown, Unranked... Unforgettable: Kentucky Basketball 1990–91, 67
UNLV Basketball Series, 77
Utilizing and Defending the Three-Point Shot, 74

War on the Boards, 78
West, Jerry, 71

Winning Basketball with Red Auerbach and Larry Bird, 78
Winning Special Situations, 75
Winning with the Flex Offense, 78
Woottan: Teaching Basketball, Vol. I, Offensive Techniques, Morgan, 75

Your Best Shot, 78

Zone Attack, 75
Zone Offense, 78

1980: That Magic Season, 70
1981: The Dynasty Renewed, 70
1982: Something to Prove, 70
1983: That Championship Feeling, 70
1984: Pride and Passion, 70
1985: Return to Glory, 70

Football

ALCOA'S Fantastic Finishes: "The Movie," 114
All the Best, 114
Alworth, Lance, 120
America's Team: The Dallas Cowboys 1975–79, 99
Auburn Football Highlights, 90
Auburn: The Decade of the Eighties, 90

Bad Boys and Good Men, 114
Beauties and the Beasts, 114
Best of the Football Follies, 123
Big Blocks and King-Size Hits, 114
Big Game America, 94
Big Plays, Best Shots and Belly Laughs, 114
Big Ten College Football Highlights 1958 and 1971, 89
Bombs Away, 115
Boom! Bang! Whap! Doink! John Madden on Football, 121
Bowden: Building a Tradition, Bobby, 90
Bradshaw, Terry, 120
Broncomania, 100

Championship Years: The Pittsburgh Steelers 1975, 1976, 1979, 1980, The, 109
Cincinnati Bengals "Stars in Stripes" 1988 Video Yearbook, 97
Coaches Video Network: One-on-One Coaching Videos, 121
Coming of Age: The Dallas Cowboys 1970–74, 99
Cotton Bowl 1990: A Big Play in Big D, 92
Crunch Course, 115
Crunchtime, 115

Defensive Line, The, 121
Denver Broncos 1989 Video Yearbook: Team Terrific, 100
Denver Broncos "Rocky Mountain Magic" 1987 Video Yearbook, 100
Distant Replay: Champions on and off the Field, 101

Era of Excellence, The 1980s, 94

Fabulous Fifties, Vol. I, The, 94
Fabulous Fifties, Vol. II, The, 94
Festival of Funnies, A, 125
Field of Honor: 100 Years of Army Football, 90
First Tastes of Glory: The Denver Broncos 1977, 1978, 1979, 1984, 100
Follies, Crunches, Highlights and Histories, 125
Football Classics, Vols., I and II, 89
Football Follies on Parade, 125
Football Follies, 125
Football Legends, 115
Football with Tom Landry: Quarterbacking to Win, 121
Foul-Ups, Fumbles and Fractures, 124

Gallant Men and Golden Moments: San Francisco's First 40 Years, 110
Giants Forever: A History of the New York Giants, 107
Gifford, Frank, 120
Gift of Grab, 115
Gigged Again: The 1990 Florida State Win over Rival Florida, 90
Glory Days of Yesterday: The Baltimore Colts 1964, 102
Golden Greats of Football, 115
Graham, Otto, 120
Great Moments in College Football, 89
Great Ones, The, 115
Great Teams/Great Years, 115
Greatest Moments in Dallas Cowboy History, The, 99
Greatest Moments of the Last 25 Years, 115
Greene, Joe, 120

Hail to the Redskins: Their First 50 Years, 112
Heart of a Champion, 115
Heritage of the Heisman Trophy, 89
High Stakes Heroes, 116

Highlights of Super Bowl I–IV 93
Highlights of Super Bowl IX–XII 94
Highlights of Super Bowl V–VIII 94
Highlights of Super Bowl XIII–XVI 94
Highlights of Super Bowl XVII–XX 94
Highlights of Super Bowl XXI–XIV 94
History of Pro Football, The, 94
Hit After Hit, 117
Hornung, Paul, 121
How to Play Winning Football, 121

In the Crunch, 116

Kicking Game, 121

Learning Football the NFL Way: Defense, 122
Learning Football the NFL Way: Offense, 122
Legacy Begins: The Miami Dolphins 1970–74, The, 104
Legend of the Lightning Bolt: History of the San Diego Chargers 110
Legendary Linemen, 116
Lombardi, 120
Lombardi: How to Play Winning Football, Vince, 122
Lombardi, Vince, 120

Mavericks and Misfits, 116
Monday Night Madness, 95
Most Memorable Games of the Decade #1, 119
Most Memorable Games of the Decade #2, 119
Most Memorable Games of the Decade #3, 119
Most Memorable Moments in Super Bowl History, 92

NFL '81, 95
NFL '87, 95
NFL Expose, 124
NFL Follies Go Hollywood, 124
NFL Head Coach: A Self Portrait, 116
NFL Playbook: A Fan's Guide to Flea Flickers, Fumbles and Fly Patterns, 116
NFL Quarterback, 116

NFL Super Duper Football Follies, 124
NFL TV Follies, The, 124
NFL's Best Ever Coaches, The, 116
NFL's Best Ever Professionals, The, 116
NFL's Best Ever Quarterbacks, The, 116
NFL's Best Ever Runners, The, 117
NFL's Best Ever Teams, The, 117
NFL's Greatest Games, Vol. II, The, 119
NFL's Greatest Games, The, 119
NFL's Greatest Hits, The, 118
NFL's Hungriest Men of the '90s, The, 118
NFL's Hungriest Men, The, 118
NFL's Inspirational Men and Moments, 118
NFL's Ultimate Football Challenge, 118
New Decade, A New Beginning, A, 90
'Noles Review: Florida State Seminoles 1990 Season Highlights, 90
Notre Dame Football 1989, 91
Notre Dame Heismans: The Men and the Moments, 91
November 10, 1979: Tennessee 40, Notre Dame 18, 91

O'Brien's Quarterback Clinic, Ken, 122
Offensive Line, The, 122
Official Pop Warner Video Handbook, 122
One Hundred Years of Volunteers Gala Celebration, 91
One Hundred Years of Volunteers, 91

Payton, Walter, 120
Playing to Win for Backs and Receivers, 122
Playing with Fire, 117
Pro Football Funnies, 125
Pro Football: The Lighter Side, 125
Professional Sports Training for Kids: Fooball with Dan Fouts, 122
Purple Power Years: The Minnesota Vikings 1969, The, 105

Quarterback Fundamentals and Techniques, 122

Rockne and the Fighting Irish, Knute, 91
Rose Bowl Highlights Through the Years, 89
Rose Parade Through the Years, 90

San Francisco 49ers "Masters of the Game" 1989 Video Yearbook, 111
San Francisco 49ers "Team of the Decade" 1988 Video Yearbook, 111
Saviors, Saints and Sinners, 95
Sayers, Gale, 120
Search and Destroy, 117
See How They Run, 117
Sensational '60s, 95
September 9, 1989: Tennessee 24, UCLA 6, 92
Shoot for the Stars, 98
Silver Celebration, 95
Simpson, O.J., 120
Son of Football Follies, 125
Sports Clinic: Football, 122
Sports Training Camp: Football, 122
Star Ascending: The Dallas Cowboys 1965–69, The, 99
Starr, Bart, 120
Starr, Bart/Johnny Unitas, 120
Strange But True Body Shapes, 117
Strange But True Football Stories, 118
Staubach, Roger, 120
Sugar Vols: Sweet Taste of Sugar and Sugar Bowl Memories, 91
Super Bowl Dream Team, 94
Super Seventies, The 95
Super Sunday: A History of the Super Bowl, 94
Superstars of the NFL, 118

Teaching Kids Football with Bo Schembechler, 122
They Wanted to Win: 1988 Notre Dame National Championship Season Highlights, 91
Three Cheers for the Redskins: The Washington Redskins 1971, 112
Three in a Row: The Green Bay Packers 1965, 1966, 101
Thunder and Destruction, 118
Tough Guys, 118

Unitas, Johnny, 120

Vintage Bo, 91
Vols 1989: Runnin' in High Cotton, 192
Vols 1990: Back on Bourbon Street, 92
Vols '87: The 4th Quarter Belongs to Us, 91

Washington Redskins "Warpath" 1987 Video Yearbook, 114
Way We Were: The New York Jets— Their First 25, The, 107
Whatever It Takes: The 1990 Texas Longhorns SWC Championship Football Season, 105
Winning Football with Vince Lombardi, 123
Winning Linebacker, The, 123
Winning Plays and Wacky Wonders, 123
Winning Tradition: The Cleveland Browns 1964, 1965, 1967, 1968, 1969, A, 98
Woody vs. Bo: The Ten Year War, 91
World Champions! Story of the 1985 Chicago Bears, 97

Years of Glory... Years of Pain: The 25-Year History of the Buffalo Bills, 96
Years to Remember: The New York Giants 1958, 1959, 107
Young, the Old and the Bold/Try and Catch the Wind, The 118

16 to 7: Alabama's Amazing 1990 Iron Bowl Victory Over Auburn, 90
1958 NFL Championship Game Highlights, 119
1960 NFL Championship Game Highlights, 119
1966 and 1967 NFL Championship Games, The, 119
1979 Chicago Bears: Go Bears!, 97

1979 Houston Oilers: Luv Ya, Blue, 113
1979 Philadelphia Eagles: The Pride of Eagle Football, 108
1979 Tampa Bay Buccaneers: From Worst to First, 112
1980 Cleveland Browns: Cardiac Kids... Again, 98
1980 Dallas Cowboys: Like a Mighty River, 99
1980 Oakland Raiders: Our Finest Hour, 103
1980 San Diego Chargers: The Power, 110
1981 & 1982 Washington Redskins: Two Years to the Title, 113
1981 Cincinnati Bengals: Stripes, 97
1981 Dallas Cowboys: Star-Spangled Cowboys, 99
1981 Miami Dolphins: Champions of the East, 104
1981 New York Giants: A Giant Step, 107
1981 New York Jets: Talk of the Town, 107
1981 San Diego Chargers: Cliffhangers, Comebacks, 110
1981 San Francisco 49ers: A Very Special Team, 110
1982 Dallas Cowboys: Great Expectations/The Man in the Funny Hat, 99
1982 Los Angeles Raiders: Commitment to Excellence '82, 103
1982 Miami Dolphins: Day of the Dolphin/NFL '82, 104
1982 New York Jets: The Road Warriors/NFL '82, 107
1982 Pittsburgh Steelers: Steel Tough Town/Steelers 50 Seasons, 109
1982 San Diego Chargers: Team of the '80s/NFL '82, 110
1983 Denver Broncos: A Team Together/NFL '83, 100
1983 Detroit Lions: Comeback Champions/NFL '83, 101
1983 Los Angeles Raiders: Just Win, Baby/NFL '83, 103
1983 Los Angeles Rams: Return of the Rams/NFL '83, 104
1983 Miami Dolphins: Day of Frustration, Season of Triumph/NFL '83, 104
1983 Pittsburgh Steelers: The Right Stuff/NFL '83, 110
1983 San Francisco 49ers: Back Among the Best/NFL '83, 110
1983 Seattle Seahawks: The Cinderella Seahawks/NFL '83, 111
1983 Washington Redskins: A Cut Above/NFL '83, 113
1984 Chicago Bears: Fight to the Finish/Road to XIX, 97
1984 Dallas Cowboys: Silver Season/Road to XIX, 99
1984 Denver Broncos: The Winning of the West/Road to XIX, 100
1984 Los Angeles Rams: A Family Tradition/Road to XIX, 104
1984 Miami Dolphins: Movers, Shakers, * Record Breakers/Road to XIX, 105
1984 New York Giants: Giants Again/Road to XIX, 107
1984 Pittsburgh Steelers: A New Beginning/Road to XIX, 109
1984 San Francisco 49ers: A Team Above All/Super Bowl XIX, 111
1984 Seattle Seahawks: One from the Heart/Road to XIX, 111
1984 Washington Redskins: Winners and Still Champions/Road to XIX, 113
1985-86 Gillette/NFL Most Valuable Player, The 115
1985 Cleveland Browns: The Division Winners, 98
1985 Dallas Cowboys: The Winning of the East, 99
1985 Los Angeles Raiders: Year of Glory, 103
1985 Miami Dolphins: Fight to the Finish, 105
1985 New England Patriots: AFC Champions, 106
1985 San Francisco 49ers: Never Surrender, 111
1986-87 Gillette/NFL Most Valuable Player, The, 115
1986 Atlanta Falcons: Falcon Fever, 96

INDEXES

1986 Buffalo Bills: Good Enough to Dream, 96
1986 Chicago Bears: We Will Be Back, 97
1986 Cincinnati Bengals: Attack, Attack, Attack, 97
1986 Cleveland Browns: Pandemonium Palace, 98
1986 Dallas Cowboys: Make Way for Tomorrow, 99
1986 Denver Broncos: Mile High Champions, 100
1986 Detroit Lions: Close Encounters, 101
1986 Green Bay Packers: A New Beginning, 101
1986 Houston Oilers: Coming of Age, 102
1986 Indianapolis Colts: A Bold New Spirit, 102
1986 Kansas City Chiefs: Flight to Prominence, 103
1986 Los Angeles Rams: Armed and Dangerous, 104
1986 Miami Dolphins: Rollercoaster Season, 105
1986 Minnesota Vikings: Back on the Attack, 105
1986 New England Patriots: Fight to the Finish, 106
1986 New Orleans Saints: Destiny in the Dome, 106
1986 New York Giants: Giants Among Men, 107
1986 New York Jets: Maximum Effort, 107
1986 Philadelphia Eagles: Coming of Age, 108
1986 Pittsburgh Steelers: Tale of Two Seasons, 109
1986 San Diego Chargers: Make Way for Tomorrow, 110
1986 San Francisco 49ers: Heart of a Champion, 111
1986 Seattle Seahawks: A Season in Three Acts, 111
1986 St. Louis Cardinals: The Right Direction, 108
1986 Tampa Bay Buccaneers: Countdown to '87, 112
1986 Washington Redskins: The Next Generation, 113
1987 Atlanta Falcons: A Preview of the '88 Falcons 96
1987 Buffalo Bills: Something to Shout About, 96
1987 Chicago Bears: Bear Down, 97
1987 Cincinnati Bengals: Great Expectations, 97
1987 Cleveland Browns: No Apologies Necessary, 98
1987 Dallas Cowboys: Blueprint for Victory, 99
1987 Denver Broncos: Champions Against All Odds, 100
1987 Detroit Lions: Working Our Way Back, 101
1987 Green Bay Packers: Pack to the Future, 101
1987 Indianapolis Colts: Off and Running, 102
1987 Kansas City Chiefs: Keeping the Faith, 103
1987 Los Angeles Raiders: Will to Win (1982–87), 103
1987 Los Angeles Rams: The Promise and the Challenge, 104
1987 Miami Dolphins: Foundation for the Future, 105
1987 Minnesota Vikings: Making a Move, 105
1987 New England Patriots: Heart of a Champion, 106
1987 New Orleans Saints: Winners, 106
1987 New York Giants: Back to the Future, 107
1987 New York Jets: Playing Hard, 108
1987 Philadelphia Eagles: Pride in Their Stride, 108
1987 Pittsburgh Steelers: Winning Ways, 109
1987 San Diego Chargers: The New Generation, 110
1987 San Francisco 49ers: One Heartbeat, 111
1987 Seattle Seahawks: Back to the Playoffs, 112
1987 St. Louis Cardinals: The 1987 St. Louis Cardinals, 108
1987 Tampa Bay Buccaneers: The Start of Something Big, 112

1987 Washington Redskins: Second to None, 113
1988 Atlanta Falcons: Fighting Falcons, 96
1988 Buffalo Bills: A Team, a Town, a Dream, 96
1988 Chicago Bears: Champions at Heart, 97
1988 Cincinnati Bengals: Men on a Mission, 98
1988 Cleveland Browns: Strange Season, 98
1988 Dallas Cowboys: A New Day in Dallas, 99
1988 Denver Broncos: A Season in Review, 100
1988 Detroit Lions: Restore the Roar, 101
1988 Green Bay Packers: Coming Together, 101
1988 Houston Oilers: Runnin', Gunnin' Excitement, 102
1988 Indianapolis Colts: A Test of Character, 102
1988 Kansas City Chiefs: Raising the Level of Expectations, 103
1988 Los Angeles Rams: One for the Books, 104
1988 Miami Dolphins: The New Generation, 105
1988 Minnesota Vikings: A Time to Step Forward, 105
1988 New England Patriots: A Team of Character, 106
1988 New Orleans Saints: Tew Tradition—Winning, 106
1988 New York Giants: The Pride Is Back, 107
1988 New York Jets: An Affirmation of Pride, 108
1988 Philadelphia Eagles: Living on the Edge, 108
1988 Phoenix Cardinals: Welcome to the NFL, Arizona, 109
1988 Pittsburgh Steelers: Forging the Future, 109
1988 San Diego Chargers: Blueprint for Glory, 110
1988 San Francisco 49ers: State of the Art, 111
1988 Seattle Seahawks: Champions of the West, 112
1988 Tampa Bay Buccaneers: Team on the Rise, 112
1988 Washington Redskins: Keeping the Faith, 113
1989 Atlanta Falcons: Special Moments—25 Years of Falcon Football, 96
1989 Buffalo Bills, 96
1989 Chicago Bears: Wounded Bears, 97
1989 Cincinnati Bengals: Unfinished Business, 98
1989 Cleveland Browns: Division Champions, 98
1989 Dallas Cowboys: Back to the Future, 100
1989 Denver Broncos: Mile High Champions, 100
1989 Detroit Lions: Foundation for the Future, 101
1989 Green Bay Packers: Out of the Pack, 102
1989 Houston Oilers: Grasping for Greatness, 102
1989 Indianapolis Colts: Hitting Full Stride, 103
1989 Kansas City Chiefs: Winning Is an Attitude, 103
1989 Los Angeles Rams: Fight to the Finish, 104
1989 Miami Dolphins: Prelude to Glory, 105
1989 Minnesota Vikings: Division Champs... Again, 105
1989 New England Patriots: Striving for Success, 106
1989 New Orleans Saints: A Fight to the Finish, 106
1989 New York Giants: Giant Achievers, 107
1989 New York Jets: A Breath of Fresh Air, 108
1989 Philadelphia Eagles: One Tough Team, 108
1989 Phoenix Cardinals: The Eldest Begins Anew, 109
1989 Pittsburgh Steelers: Yes, We Can! 109
1989 San Diego Chargers: Re-Charged for the '90s, 110

1989 San Francisco 49ers: Back to Back, 111
1989 Seattle Seahawks: Northwest Passage, 112
1989 Tampa Bay Buccaneers: We're Buccaneers, 112
1989 Washington Redskins: Back on the Warpath, 113
1990 Atlanta Falcons, 96
1990 Buffalo Bills, 96
1990 Chicago Bears, 97
1990 Cincinnati Bengals, 98
1990 Cleveland Browns, 98
1990 Dallas Cowboys, 100
1990 Denver Broncos, 101
1990 Detroit Lions, 101
1990 Green Bay Packers, 102
1990 Houston Oilers, 102
1990 Indianapolis Colts, 103
1990 Kansas City Chiefs, 103
1990 Los Angeles Raiders, 104
1990 Los Angeles Rams, 104
1990 Miami Dolphins, 105
1990 Minnesota Vikings, 105
1990 New England Patriots, 106
1990 New York Giants, 107
1990 New York Jets, 108
1990 Philadelphia Eagles, 108
1990 Phoenix Cardinals, 109
1990 Pittsburgh Steelers, 109
1990 San Diego Chargers, 110
1990 San Francisco 49ers, 111
1990 Seattle Seahawks, 112
1990 Tampa Bay Buccaneers, 112
1990 Washington Redskins, 113
1991 Atlanta Falcons, 96
1991 Buffalo Bills, 97
1991 Chicago Bears, 97
1991 Cincinnati Bengals, 98
1991 Cleveland Browns, 98
1991 Dallas Cowboys, 100
1991 Denver Broncos, 101
1991 Green Bay Packers, 102
1991 Houston Oilers, 102
1991 Indianapolis Colts, 103
1991 Kansas City Chiefs, 103
1991 Los Angeles Raiders, 104
1991 Los Angeles Rams, 104
1991 Miami Dolphins, 105
1991 Minnesota Vikings, 106
1991 New England Patriots, 106
1991 New Orleans Saints, 106
1991 New York Giants, 107
1991 New York Jets, 108
1991 Philadelphia Eagles, 108
1991 Phoenix Cardinals, 109
1991 Pittsburgh Steelers, 109
1991 San Diego Chargers, 110
1991 San Francisco 49ers, 111
1991 Seattle Seahawks, 112
1991 Tampa Bay Buccaneers, 112
1991 Washington Redskins, 113
25 Years of Georgia Football: The Vince Dooley Era, 90

Hockey

Blades of Summer, 128
Broten's Gold Medal Hockey, Neal, 131

C of Champions, 128
Canada/Russia Games 1972, 128
Cherry's Rock 'Em Sock 'Em Hockey, Don, 128
Clarke, Bobby, 130

Devastating Hits in Hockey, 128
Dynamite on Ice, 128

Esposito, Phil, 130

Fantastic Hockey Fights, 128

Great Plays From Great Games, 129
Gretzky: Above and Beyond, Wayne, 130
Gretzky: Hockey My Way, Wayne, 131

Hilarious Hockey Highlights, 132
Hockey for Kids and Coaches, 131
Hockey Slapshots & Snapshots, 129
Hockey Super Stars. Orr, Hull, Howe, Esposito, 130
Hockey: A Brutal Game, 129
Hockey: Here's Howe, Conditioning and Coaching, 131
Hockey: Here's Howe, Defense, 131
Hockey: Here's Howe, Forwards, 131
Hockey: Here's Howe, Goaltending, 131
Hockey: Here's Howe, Power Skating, 131

Hockey: Here's Howe, Shooting, 131
Hockey: Here's Howe, Stick Handling and Passing, 131
Hockey: The Lighter Side, 132
Hockey's Greatest Hits, 129
Hockey's Hardest Hitters, 129
Howe, Gordie, 130
Hull, Bobby, 130

Lemieux: Success Story, Mario, 130
Les Canadiens, 129
Lightning on Ice: The History of Hockey, 129
Los Angeles Kings: 1988–89, 129
Los Angeles Kings: 1989–90, 129

Mario the Magnificent, 130

New York Rangers Highlights 1989–90, 129

Orr, Bobby, 130

Pro Hockey Funnies, 132

Rangers, Story of the 88–89: Year of the Rookies, 129
Rough and Tough Hockey: Don Cherry Vol. 2, 129

Sports Champions: Wayne Gretzky, 130
Stanley Cup Playoffs, 129
Super Dooper Hockey Bloopers, 132

Time to Remember, A, 129
Tradition on Ice: 62 Year History of the New York Rangers, 129

Soccer

Attack, The, 138

Ball Control, 139
Building a Relationship With Your Soccer Ball, 137

Coaching Youth Soccer: Official U.S. Youth Soccer Association Coaching Guide, 137
Creating Space, 138

Defending, 138
Developing Fast Feet/Shielding the Ball, 137
Do It Better: Soccer, 137
Dribbling and Feinting, 139

Goalkeeping, 138
Goals Galore: Over 100 of Soccer's Greatest Goals, 136
Graduated Soccer Method Series: Fundamentals and Techniques, 137

Head to Toe: Soccer for Little Leaguers, 137
Hot Tips Soccer Series: Bobby Charlton's Soccer, Level 1, 137
Hot Tips Soccer Series: Bobby Charlton's Soccer, Level 2, 137

Kicking, 139

Passing and Support, 138
Pélé, 136
Pélé: The Master and His Method, 137

Set Plays, 138
Shooting, 138
Soccer Fundamentals, 137
Soccer Series: Basic Individual Skills, Offensive and Defensive Maneuvering, Goal Keeping, 138
Soccer Tactics and Skills, 138
Soccer with the Superstars, 138
Soccerobics, 139
Sports Clinic: Soccer, 138
Sports Training Camp: Soccer, 138
Striker Tactics: Skills to Help You Score, Part I, 138
Striker Tactics: Skills to Help You Score, Part II, 138
Super Skills and Heading, 139

Taking on Your Opponent, 137
Teaching Kids Soccer with Bob Gansler, 138

Videocoach Vogelsinger's Soccer Series, 139

Winning Soccer: Basics of the Game, 139

Boxing

Ali-Foreman: The Rumble in the Jungle, 145
Ali-Frazier: The Thrilla in Manilla, 145
Ali-Norton Trilogy, 145
Ali: Boxing's Best, Muhammad, 145
Ali, Muhammad, 145
Ali vs. Zora Folley: March 22, 1967, Muhammad, 145

Baer vs. Louis 1935 / Louis vs. Schmeling 1936, 143
Boxing's Greatest Champions, 143
Boxing's Greatest Knockouts and Highlights, 143
Boxing's Greatest Knockouts, Vol. I, 143
Boxing's Greatest Upsets 143

Champions Forever, 144
Clay-Liston: Clay Shocks the World, 145

Dempsey, Jack 145
Douglas-Tyson: The Upset of the Century, 146

Fabulous Four, The, 144
Frazier, Joe, 146

Greatest Rounds Ever, Part 1, 144
Griffith's Learn to Box, Emile, 146
Grudge Fights, 144

Heavyweights: The Big Punchers, The, 144

Heavyweights: The Stylists, The, 144

Johnson, Jack, 146

Legendary Champions, 144
Legends of the Ring, 144
Leonard-Duran III: "Uno Mas", 146
Leonard vs. Hearns Saga, The, 146
Louis, Joe, 146
Louis: Fort All Time, Joe, 146

Marciano, Rocky, 146
Middleweights, 144

Power Profiles: Jack Dempsey, 146
Power Profiles: Joe Louis, 146

Ringside with Mike Tyson, 144
Robinson, Sugar Ray, 146

Sports Champions: Sugar Ray Leonard, 146

Ten and Counting: Best of ESPN Boxing, 144
Tyson and History's Greatest Knockouts, Mike, 145
Tyson and the Heavyweights, Mike, 145
Tyson's Greatest Hits 144

30 Great One-Punch Knockouts, 144
30 More One-Punch Knockouts, 144

Tennis

Attack, 150

Best of U.S. Open Tennis, 149
Body Prep, Tennis!, 150
Borg, Bjorg, 151
Braden Tennis, Vic, 150

Connors, Jimmy, 151
Connors: Match Strategy, Jimmy, 150
Connors' Tennis: Winning Fundamentals, Jimmy, 150

Golden Greats of Tennis, 149

McEnroe and Ivan Lendl: The Winning Edge, John, 150
Maximizing Your Tennis Potential with Vic Braden, 150

Official 1989 U.S. Open Tennis Video, The, 149

Smash Hit, 149
Smashing Ladies: The Legends of Women's Tennis, 149

Teaching Kids Tennis, 150
Tennis From the Pros, 151
Tennis Our Way, 151
Tennis to Win, 151
Tennis with Van der Meer, 151

Visual Tennis, 151

Wade's Class, Virginia, 151
Warm Up to Attack, 151
Wimbledon: The One to Win, 149

1988 U.S. Open Video, The, 149

Golf

Ace of Clubs, 174
Alcott: Winning at Golf, Amy 170
Alliss: Play Better Golf, Peter, 159
An Inside Look at the Game for a Lifetime, 162
Approach & Sand Play, 166
Approach Game, The, 171
Armour: How to Play Your Best Golf All the Time, Tommy, 159
Armstrong: A Picture's Worth 1000 Words, Wally, 159
Armstrong: Feel Your Way to Better Golf, Wally, 159
Armstrong: Golf Gadgets & Gimmicks, Wally, 159
Armstrong's Collection of Teaching Aids and Drills, Wally, 159
Armstrong's Golf for Kids of All Ages, Wally, 171
Armstrong's Golf: The Easy Way, Wally, 159
Azinger on Fairway and Green Sand Traps, Peter, 164
Azinger Way, The, 159

Ballard: Fundamental Golf Swing, Jimmy, 159
Ballesteros: A Study of a Legend, Seve, 158
Barber on the Driver and Wedge, Miller, 171
Bay Course At Kapulua, 156
Best of Bobby Jones, The, 163
Biomechanics of Power Golf, 160
Birdies and Bloopers, 174
Brainwaves Golf, 160
Bullybunion: The Old Course, 156

Carner's Keys to Great Golf, Joanne, 170
Casper: Golf Basics, Billy, 160
Casper: Golf Like a Pro, Billy, 160
Casper: Secrets of Golf, Billy, 160
Charles: Golf From the Other Side, Bob, 160
Chi Chi's Bag of Tricks: In and Out of Trouble With Chi Chi Rodriguez, 160

Chipping and Putting Video, The, 160
Classic Golf Experiences, 156
Clubs, The, 166
Control Your Swing Plane, 167
Couples on Tempo, Fred, 164
Course Strategy, 166
Crampton: Total Golf, Bruce, 160
Crenshaw: The Art of Putting, Ben, 160
Cullen: Hit It Farther, Betsy, 170
Cures for Crooked Shots, 167

Daly's... the Long Shot: 1991 PGA Championships, 154
Disney World: Magnolia, Walt, 158
Developing Maximum Consistency, 163
Doral's Blue Monster, 157
Dorf on Golf, 174
Dorf's Golf Bible, 175
Douglass: Rhythm, Tempo, Sand, and Chip Shots, Dale, 171
Drive for Distance, 161
Dunaway: Power Driving, Mike, 160

Earp's Golf Lesson, Charlie, 160
Effortless Power, 167
Etiquette of Golf, 160
Exercise Fitness for Golf Series, 173
Exercise Fitness for Golf, 173
Exercises for Better Golf, 173

Fabulous Finishes of the PGA Tour, 154
Fabulous Putting, 154
Faldo is Golf, Nick, 161
Faldo's Golf Course, Nick, 161
Fantastic Approaches: The Pro's Edge, 161
Find Your Own Fundamentals, 161
Fit for Golf, 174
Floyd: 60 Yards In, Ray, 161
Four Absolutes, The, 161
From Fairway to Green, 165
From Tee to Fairway, 165
From Tee to Green, 163
Full Swing, The, 170

Full Swing, The, 163

Geiberger: 6-in-1 Golf Clinic, Al, 161
Geiberger: Golf, Al, 161
Geiberger: Winning Golf, Al, 161
Golden Greats of Golf, 154
Golf Comedy Classics, 175
Golf Digest Video Almanac 1989, 154
Golf Digest Videos, 161
Golf for Winners, 162
Golf Fundamentals: The Ben Sutton Golf School, 162
Golf Funnies, 175
Golf Swing, 166
Golf Tips from 27 Top Pros, 162
Golf Trick Shots, 175
Golf with the Super Pros, 162
Golfbusters: The Lighter Side of Golf, 175
Golflex, 174
Golf's Gambling Games, 175
Golf's Goof-Ups & Miraculous Moments, 175
Golf's Greatest Moments, 155
Golf's Greats: Vol. I, 155
Golf's Mental Keys, 167
Golf's One in a Million Shots, 155
Good Grief! Golf?, 175
Great Golf Courses of Southern California, 157
Great Golf Courses of the World: Scotland, 157
Great Moments of the Masters, 155
Greatest 18 Holes of Championship Golf, 157
Grip, Stance, Aim & Posture, 167
Grout: Keys to Consistency, Jack, 162
Grout: The Last 100 Yards, Jack, 162

Haines: Golf Etiquette, John, 162
Hamm: Hit It Long, Jack, 162
Harbour Town, 157
History and Traditions of Golf In Scotland, 157
History of the PGA Tour, 155
History of the Ryder Cup, 155
Hitting the Long Shots, 162
How to Break 90 in 30 Days, 162
How to Win at Golf Without Really Cheating, 175

Imagine All Eagles, 155
Irons: Hitting with Power and Accuracy, 168
Irwin: Difficult Shots, Hale 162

Jacobs: Faults and Cures, John, 163
Jacobs: Full Swing, John, 163
Jacobs: Short Game, John, 163
Jones Instructional Series, Bobby, 163
Jones Limited Collector's Edition, Bobby, 163
Just Missed, Dammit!, 175

Killer Golf: 10 Tips to Take 10 Strokes Off Your Game, 163
Kite: Reaching Your Golf Potential, Tom, 163
Knudson: The Swing Motion, George, 163
Koch on Putting, Gary, 164

La Costa, 157
Ladies, Effortless Power, 167
Leadbetter: The Full Golf Swing, David, 163
Leadbetter: The Short Game, David, 164
Linton: Better Golf with a Little Bit of Magic, Bill, 164
Little Scams on Golf, 175
Littler: The 10 Basics, Gene, 164
Lopez: Golf Made Easy, Nancy, 170
Love III on Driving, Davis, 164

McTeigue: Keys to the Effortless Golf Swing, Michael, 165
Mann: Let'S Get Started, Bob 164
Mann's Automatic Golf: The Method, Bob, 164
Mann's Complete Automatic Golf Method, Bob, 164
Mann's Golf: The Specialty Shots, Bob, 164
Master System to Better Golf, The, 164
Mastering the Basics, 170
Mastering the Fundamentals, 166
Masters, The 156
Masters Tournament 1986: A Golden Moment In the History of Sports, 155

Masters Tournament 1987, 155
Masters Tournament 1988, 155
Masters Tournament 1989, 155
Masters Tournament 1990, 155
Masters Tournament 1991, 155
Mid Irons & Short Irons, 166
Miller: Golf the Miller Way, Johnny, 165
Moody: Long Irons and Putting, Orville, 171
Morris: Keeper of the Greens, Tom, 158
Murphy's Laws of Golf, 175
Murray: Power Driving, Kelly, 165
Myrtle Beach Golf: Great Golf Courses You Can Play!, 157

Nice Shot!, 165
Nicklaus' Golf My Way I: Hitting the Shots, Jack, 165
Nicklaus' Golf My Way II: Playing the Game, Jack, 165
Nicklaus' Golf My Way: the Full Swing, Jack, 165
Norman: Complete Golfer, Greg, 165

One Club Challenge, 175
One Move to Better Golf, 165
Outrageously Funny Golf, 176
Overton: 50 Plus Senior Golf, Jay, 171

PGA Tour Golf 1: The Full Swing, 166
PGA Tour Golf 2: The Short Game, 166
PGA Tour Golf 3: Course Strategy, 166
PGA Tour Golf, 166
PGA Tour: 18 Toughest Holes, 157
PGA West Stadium Course, 157
Palmer Story, Arnold, 158
Palmer, Arnold, 158
Palmer: Play Great Golf, Arnold, 166
Pitching & Sand Play, 168
Pitching, Chipping, Putting, 167
Play Senior Golf, 171
Play Your Best Golf, 166
Player on Golf, Gary, 166
Practice Like a Pro, 166
Purtzer on Iron Accuracy, Tom, 164
Putting & Chipping, 166, 168

Putting for Profit, 161
Putting with Confidence, 167
Putting: The Science and the Stroke, 168

Rankin's Break 90 In 21 Days, Judy, 170
Rhodes: Get Fit for Golf, John, 174
Ritson: Golf Your Way, Phil, 167
Ritson: Video Encyclopedia of Golf, Phil, 167
Rogers: The Short Game, Phil, 167
Rosburg: Golf Tips, Bob, 167
Rosburg's Break 90 in 21 Days, Bob, 167
Rules of Golf, 167
Runyon: The Short Way to Lower Scoring, Paul, 167
Ryder Cup 1991, 156

St. Andrews Old Course, 157
Sand Magic, 167
Save from the Sand, 161
Saving Strokes with the Rules, 168
Schlee: Mastering the Long Putter, John, 168
Schlee: Maximum Golf, John, 168
Scholastic Golf, 171
Scoring Zone, The, 166
Secrets of the Power Fade, 167
Sellinger: Drive My Way, Art, 168
Seniors, 171
Seniors, Effortless Power, 167
Sharpen Your Short Irons, 161
Sharp's Golf: I Hate This Game!, Thom, 176
Sheehan: Sybervision Women'S Golf, Patty, 170
Snead, Sam, 158
Snead's Secrets for Seniors, Sam, 171
Sports Psychology: the Winning Edge in Sports, 174
Stadler on the Short Game, Craig, 164
Stephenson's How to Golf, Jan, 170
Stockton: Precision Putting, Dave, 168
Stockton's Golf Clinic, Dave, 168
Strange: How to Win and Win Again, Curtis, 168

Strategies & Skills, 166
Strategies & Techniques, 163
Strategies, The, 166
Stretching for Better Golf, 174
Stretching: The Driving Force, 174
Suntory World Match Play, 156
Super Golf for Juniors, 171
Swing for a Lifetime, A, 161

TPC at Sawgrass, 157
Teaching Kids Golf, 172
Ten Years of the British Open: The 1980s, 156
Three Men and a Bogey, 176
Tiger Shark Video Library, 168
Tips from the Tour, Volume 1, 168
Toski Teaches You Golf, Bob, 168
Toughest 18 Holes of Golf in America, 157
Trevino, Lee, 158
Trevino's Golf Tips for Youngsters, Lee, 172
Trevino's Legacy of the Links, Lee, 157
Trevino's Priceless Golf Tips, Lee, 169
Trevino's Putt for Dough, Lee, 169
Trouble Shots: Recovery Strokes Made Easy, 168
Trouble Shots; Great Escapes, 162

U.S. Open 1987, 156
U.S. Open 1988, 156
U.S. Open: Golf's Greatest Championship, 156
U.S. Women'S Open 1988: A Star is Born, 156
Using the Wind to Win, 167

Venturi: Better Golf Now!, Ken, 169
Venturi: Stroke Savers, Ken, 169

Wadkins on Trouble Shots, Bobby, 164
Weiskopf: the Golf Swing, Tom, 169
When the Chips are Down. 162
White: Golf for Women, Donna, 170
White's Beginning Golf for Women, Donna, 170
Wide World of Golf: Video Magazine, 172
Winning Pitch Shots, 162
Wiren: Super Power Golf, Dr. Gary, 174
Wiren: The Greater Golfer in You, Dr. Gary, 169
Woman's Golf Fundamentals: The Ben Sutton Golf School, 170
Women's Golf Instructional Series, 170
Woods & Long Irons, 166
Woods: Hitting for Distance, 168
Woosnam: Power Game, Ian, 169
World's Worst Avid Golfer, 176

Zarley: Golf for All Age Groups, Kermit, 169
Zoeller: Scramble to Better Golf, Fuzzy, 169

10 Fundamentals of the Modern Golf Swing, 168
101 Great Putts, 156
1750 Masters Tournament, The, 155
25 Great Pros' Second Shots, 169
9 Tips From 9 Legends of Golf, 165

General Olympics and Other Sports

Advanced Fastball Pitching, 191
Afternoon With Gerry Watson: Fundamental Techniques and Professional Secrets (Billiards), An, 187
All New Bob Uecker's Wacky World of Sports, 193
All New Not-So-Great Moments in Sports, 193
Amazing Bif Bam Boom Anything Goes Sports Bloopers, 193
Andretti, Mario, 181
Anthony: Going for 300, Earl, 187
Archery Series: Invitation to Archery, 187

Basic Skills and Better Techniques, 193
Berman Unbelievable Sports Plays, Chris, 181
Best of ABC's Wide World of Sports: The '60s, 181
Best of ABC's Wide World of Sports: The '70s, 181
Best of ABC's Wide World of Sports: The '80s, 181
Best of Bob Uecker's Wacky World of Sports, 193
Best of Comedy, Volume 1, The, 194
Bizarre Sports and Incredible Feats, 181, 194
Body Prep: The Ultimate Ski Fitness Video, 189
Bogner's Skiing Techniques, Peter, 190
Bowl to Win with Earl Anthony, 187
Bowling Series: New Approach to a Great Old Game, 187
Bowling with Nelson Burton, Jr., 187
Breakthrough Basics of Downhill Skiing with Hank Kashiwa, 190
Brine Lacrosse Clinic, 188
Byrne's Standard Video of Pool and Billiards, Vol. II, 187
Byrne's Standard Video of Pool and Billiards, 187

Championship Softball Hitting System, 191
Championship Video Series: Sprints And Relays, 189
Coach to Win: Coaching Middle School And Elementary Volleyball, 192
Coe: Born to Run, Sebastian, 189
Cross-Country Skiing Basics, 190

Defensive Fundamentals and Drills, 191
DeVarona'S No Impact Workout: Swim Your Way to Fitness, Donna, 192
Distinctive Skiing, 190
Diving My Way, 192
Do It Better: Beach Volleyball, 193
Do It Better: Running, 189
Do It Better: Slo Pitch Softball, 191
Do It Better: Volleyball, 193
Do It Better: Wrestling, 193
Dorf and the First Games of Mt. Olympus, 194
Downhill Skiing Basics, 190
Downhill Skiing Primer, The, 190

Elliott: Racing into History, Bill, 181
Escape to Ski, 182

Fast Pitch Team Defense, Strategy And Sliding, 191

Gambril's Classic Series, Don, 192
Get the Feeling: Power, 182
Get the Feeling: Speed, 182
Get the Feeling: Winning, 182
Gold Medal Champions, 182
Golden Goofy Classics, 194
Great Sports Moments of the '80s, 182
Greatest Moments in Chicago Sports History, 182
Greatest Moments in Philadelphia Sports History, 182
Greatest Sports Follies, 194

Greatest Sports Legends 10th Anniversary Special: 101 Superstars, 183
Greatest Sports Legends: 160 Superstars of the 20th Century, 183
Greatest Sports Legends: Hall of Fame, 182
Greatest Sports Legends: Miracle Moments of Sport, 182
Greatest Sports Legends: Record Breakers, Volume I, 183
Gymnastics Fun with Bela Karoli, 188

Hard Road to Glory, 183
Heroes and Heartaches: A Treasury of Boston Sports Since 1975, 183
Heroines: Early Women Sports Stars, 183
Highlights of the 1988 Summer Olympics (Seoul), 183
History of the Indianapolis 500, The, 183
Hockey: The Basic Skills, 188
Horses Talk: The Paddock and Post Parade, 188
How to Ice Skate, 188
How to Play Pool with Minnesota Fats, 187

Ice Skating Showcase: Great Routines of the '80s, 183

Jewels of the Triple Crown, 183

Keys to Fitness, Nutrition, and Safety, 193
Kidd Ski Racing: The Fast Way to Improve Your Skiing, Billy, 189
Kids Wrestling Series, 193

Let's Bowl with Dick Weber, 187
Live and Drive the Indy 500, 183

Magic Memories on Ice, 183
Magicians of Sport, 184
Men's Swimming, 192
Miller's Extreme Skiing, Warren, 186
Miller's Learn to Ski Better, Warren, 190
Miller's Ski Country, Warren, 186
Miller's Ski Time: A Ski Vacation in a Box, Warren, 186
Miller's Snowonder, Warren, 186
Miller's Sports Bloopers, Warren, 194
Miller's Steep and Deep, Warren, 186
Miller's Steeps, Leaps and Powder, Warren, 186
Miller's White Winter Heat, Warren, 186

Not-So-Great Moments in Sports, Take 3, The, 194
Not-So-Great Moments in Sports, The, 194

Official 1988 Calgary Winter Olympics Video, 184
Olympic Boxing, 184
Olympic Gold Medal Winners: The First 90 Years, 184
Olympic Gymnastics, 184
Olympic Track & Field (Men), 184
Olympic Track & Field (Women), 184
Olympic Volleyball, 184
Olympic Water Sports, 184
Organizing a Kids' Wrestling Club, 193
Owens, Jesse, 183

Petty, Richard, 185
Pitching Slow Pitch Softball, 191
Power Hitting in Softball, 191
Pro Sports Bloopers, 194

Rare Sportsfilms (RSF) Auto Racing, 184
Record Breakers of Sport, 185
Rock Chalk, Jayhawk, 185
Running Great with Grete Weitz, 189

Score More Bowling with Nelson Burton, Jr., 187
Skating Away: Cross-Country Skiing, 190
Ski the Mahre Way, 190
Skiing Extreme: Hot Music! Radical Runs! Big Air!, 185
Skiing with Style: Mastering the Mountain, 190

Slo Pitch Softball: Reflex Hitting System with Ray DeMarini, 191
Slow Pitch Team Defense, Strategy and Sliding, 191
Softball Series: Basic Skills in Softball, 191
Softball: Putting it Together, 191
Speed and Explosion, 189
Sports Bloopermania, 194
Sports Bloopers Awards with Chris Berman, 194
Sports Colossus: Heroes of the '20s, '30s, and '40s, 185
Sports Heroes, 186
Sports IQ Test, 186
Sprinting with Carl Lewis and Coach Tom Tellez, 189
Strategy of Pitching Slo Pitch Softball, 191
Strike: The Guide to Consistent Bowling with Joe Berardi, 188
Super Sports Follies, 194
Swim Lessons for Kids: A Simple, Proven Method for Parents to Teach Their Children over Three to Swim, 192
Swim Smarter, Swim Faster, 192
Swimming: Excellence in Swimming Stroke Technique, 192

Teaching Kids Bowling, 188
Teaching Kids Skiing, 190
Teaching Kids Speed for all Sports with Carl Lewis, 189
Teaching Kids Swimming with John Naber, 192
Time Capsule: Los Angeles Olympic Games of 1932, 186
Time Out for Hilarious Sports Bloopers, 194

Time Waits for Snowman, 186
Track & Field: Coaching by the Expert Olympic Team Coaches Series, 189
Triathlon Training and Racing with Dave Scott, 192

Water is Friendly: First Step in Learning to Swim, 192
Women's Swimming, 192
World Wide Sports Bloopers, 194

16 Days of Glory II, 185
16 Days of Glory, 185
1930s-1964 "News From Indy", 184
1949 Indy "Behind the Checkered Flag", 184
1951-59 "Pioneers of Stock Car Racing", 184
1952 Daytona, 185
1952 Indy "The Fabulous 500", 184
1953 Indy "The Hottest 500", 184
1954 Indy "The Fantastic 500", 184
1956 "Racing Tripleheader", 184
1956 Southern 500 at Darlington, 185
1958 Daytona, 185
1961 Southern 500 at Darlington, S.C., 185
1963 Atlanta 500 at Atlanta International Raceway, 185
1963 Riverside 500, 185
1963 Southern 500 at Darlington, 185
1964 Indy "Way of a Champion", 184
1970 Grand National Highlights, 185
1971 Southern 500 at Darlington, 185
25 Years of Sports: 1943-67, 186

Movies

Alibi Ike, 196
All-American, The, 208
All the Right Moves, 208
Amazing Grace and Chuck, 206
Angels in the Outfield, 196
Aunt Mary, 197

Babe, 224
Babe Ruth Story, The, 197
Bad News Bears, The, 197
Bad News Bears Go to Japan, The, 197
Bad News Bears in Breaking Training, The, 197
Bang the Drum Slowly, 198
Best of Times, The, 208
Bingo Long Traveling All-Stars and Motor Kings, The, 198
Body and Soul, 215
Brian's Song, 209
Bull Durham, 198

Champ, The, 216
Champ, The, 216
Champion, 216
Champions, 223
Chariots of Fire, 224
Coach, 206
Color of Money, The, 223
Comeback Kid, The, 199
Crazylegs, 209

Damn Yankees, 199
Dead Solid Perfect, 220
Deadliest Season, The, 214
Downhill Racer, 221

Easy Living, 209
Eight Men Out, 199
Elmer the Great, 199
Everybody's All-American, 209

Fat City, 216
Father Was a Fullback, 210
Fear Strikes Out, 200
Field of Dreams, 200
Fighting Back, 210
Fish That Saved Pittsburgh, The, 206

Flesh and Blood, 216
Follow The Sun, 221
Freshman, The, 210

Go, Man, Go, 207
Going for the Gold: The Bill Johnson Story, 221
Golden Boy, 216
Golden Gloves Story, The, 217
Grambling's White Tiger, 210
Great American Pastime, The, 200
Great White Hope, The, 217

Harder They Fall, The, 217
Heart of a Champion: The Ray Mancini Story, 217
Hoosiers, 207
Horse Feathers, 210
Hustler, The, 222

Iron Major, The, 211
Iron Man, 217
It Happens Every Spring, 200
It's Good to be Alive, 201

Jocks, 222
Johnson, Jack, 218

Kid From Brooklyn, The, 218
Kid From Cleveland, The, 201
Kid From Left Field, The, 201
Kill the Umpire, 202

Little Mo, 222
Loneliness of the Long Distance Runner, The, 224
Long Gone, 202
Longest Yard, The, 211
Louis Story, The Joe, 218
Love Affair: The Eleanor and Lou Gehrig Story, A, 202

Major League, 202
Make Mine Music, 202
Mathias Story, The Bob, 224

Nadia, 225
Natural, The, 203

North Dallas Forty, 212
Number One, 212

One in a Million: The Ron LeFlore
   Story, 203
One on One, 207
Other Side of the Mountain, 222
Owens Story, The Jesse, 224

Pat and Mike, 221
Personal Best, 225
Phar Lap, 223
Pride of St. Louis, The, 203
Pride of the Yankees, The, 203
Prize Fighter, The, 218

Raging Bull, 218
Requiem for a Heavyweight, 218
Rhubarb, 204
Robinson Story, The Jackie, 201
Rockne: All-American, Knute, 211
Rocky, 219
Rocky II, 219
Rocky III, 219
Rocky IV, 219
Rocky V, 219
Rose Bowl Story, The, 212

Safe at Home, 204

Saturday's Hero, 212
Saturday's Heroes, 212
See How She Runs, 225
Semi-Tough, 212
Set-Up, The, 219
Slap Shot, 214
Slugger's Wife, The, 204
Somebody Up There Likes Me, 220
Spirit of West Point, The, 213
Stealing Home, 204
Stratton Story, The, 205

Take Me Out to the Ball Game, 205
That Championship Season, 207
That's My Boy, 213
Thorpe: All-American, Jim, 211
Tiger Town, 205
Triumph of the Spirit, 220
Trouble Along the Way, 213

Victory, 215

What Price Victory, 213
Winner Never Quits, A, 205
Winning Team, The, 205
World in my Corner, 220

Youngblood, 214